Advances in Cereals Processing Technologies

NEW INDIA PUBLISHING AGENCY
New Delhi-110 034

Advances in Cereals Processing Technologies

Editor-in-Chief

Gopal Kumar Sharma FAFST(I)

Former Additional Director & Head
Grain Science and Technology Division
DRDO-DFRL (Defence Food Research Laboratory)
Ministry of Defence, Government of India Mysuru,
Karnataka

Edited by

Anil Dutt Semwal

Director
DRDO-DFRL (Defence Food Research Laboratory)
Ministry of Defence, Government of India Mysuru,
Karnataka

Dev Kumar Yadav

Scientist 'D'
Grain Science and Technology Division
DFRL-DRDO (Defence Food Research Laboratory)
Ministry of Defence, Government of India Mysuru,
Karnataka

CRC Press is an imprint of the
Taylor & Francis Group, an informa business

NEW INDIA PUBLISHING AGENCY

New Delhi-110 034

First published 2022
by CRC Press
4 Park Square, Milton Park, Abingdon, Oxon, OX14 4RN

and by CRC Press
6000 Broken Sound Parkway NW, Suite 300, Boca Raton, FL 33487-2742

CRC Press is an imprint of Taylor & Francis Group, an Informa business

British Library Cataloguing-in-Publication Data
A catalogue record for this book is available from the British Library

Library of Congress Cataloging-in-Publication Data
A catalog record has been requested

ISBN: 978-1-032-19845-3 (hbk)
ISBN: 978-1-003-26112-4 (ebk)

DOI: 10.1201/9781003261124

Preface

Developing countries including India emphasise for the trained manpower with fundamental and advanced understanding of science and technology specially engaged in food processing and R&D sector. These trained human resources basically control production and maintain quality of the finished products which largely affects the consumer's acceptability so the market. In India immense emphasis was laid on the quality attributes of cereals and cereal based product owing their lion's share in food processing sector, be it bakery, brewery milling, snacks or sweet sector. Cereal grains have represented the principal component of the human diet since long. Their processing comprises an important part of the food production chain, but it is a complex procedure. This ensures their value addition in terms of quality, availability, cost and acceptability. Since cereals are an important source of carbohydrates, proteins, lipids, vitamins (mainly of B-complex) and vitamin E, and inorganic and trace elements, the reutilization and valorisation of their by-products is a great challenge toward the sustainable development of the agro-food sector. The present compilation is the results of sincere efforts of the contributing authors and editorial board members to present its reader with comprehensive knowledge related to cereals processing. It is imperative to ahave sound knowledge of Food laws and regulations with an Indian perspective as these plays a pivotal role in commercializing food products as well as fresh produce which are aptly covered in this book. It includes recent trends in technology of cereals based products, technological updates in legumes and pulses based convenience/ processed foods, various aspects of evolution of bakery and confectionery technology, Technological evaluation of milling. Since age's process of fermentation was employed for preserving the cereals based food by using general and specified micro flora and micro fauna. The science and technology involved is well explained in chapter titled Fermented foods based on cereals and pulses. The most important quality attributes related to cereals processing are rheological and thermal changes which occur when extrinsic factors such as moisture and temperature are ebbed and flowed. This subject was sensibly covered under Rheological & thermal changes occurring during processing. Sugarcane and sugar industry have the largest contribution to the industrial development. Rural economy and livelihood is mainly based on sugarcane and sugar industry as sugarcane is the cash crop for most of the farmers. Various unit operations and technology involved are explained as Recent updates in sugar, honey, jaggery and salt processing. One of the most important requirements of the commercial market is to have good shelf life stability of the products with respect to various chemical parameters attributed to the oxidative changes in processed Foods and is also aptly covered here. As the country is at the threshold of emerging as a dominant global player in food processing, the present book is expected to contribute adequately in ensuring development of competent technical manpower to support both industry and academia.

We sincerely hope that valuable knowledge gathered by authors and writing chapter in their renowned filed will bring revolutionary changes and fulfil the expectations of readers.

Editor-in-Chief

Contents

1

Food Laws and Regulatory Authorities An Indian Perspective

***Lovepreet Kaur*[1], *Ajay Singh*[1], *Pankaj Kumar*[2], *Kamalpreet Singh*[1] and *Damanjeet Kaur*[2]**

[1]Department of Food Technology, Mata Gujri College, Fatehgarh Sahib, Punjab
[2]Department of Microbiology, Dolphin (PG) Institute of Biomedical and Natural Sciences, Dehradhun, Uttarakhand

Introduction

Food is the basis to sustain a healthy life as evidenced from its consumption through centuries by human. There are plenty of plenty of foodstuffs available these days as part of new product development and flexible global participation of researchers and food technocrats. Their researches brings a lot reforms in traditional recipes through newest technological concepts (functional or nutraceutical) that address much of safety aspect for consumers health and is the prime focused concern for now (*Keservani, et al 2014).* Even than strict labeling policies is the present hour need to stay fit and fine. Safe and wholesome supply of foods, high pace food demands, increase in processing aspect, minimization of fraudulent practices are the key points that somehow require laws and standards during transportation, storage, marketing and storage.

Nation's growth resides in their healthy people that contribute flawless to acquire achievements in overall growth of country. So availability of healthy food is must which can only be governed through strict standards and regulatory agency's regulation. Journey of food from "**Production to consumption**", "**Farm to fork**", "**Unstable to table**" and "**Boat to throat**" are the activities where these agencies abide their regulations by way of laws and standards. Indian market somehow lacking its policy to organize food regulation appropriately hence is segmented in two types of food products *viz*;

- Standardized Food Products: For standardized food products, specific standards are prescribed and they do not require product approval before manufacturing, sale and distribution process. Example: Fruit Drinks, Biscuits and Sauces etc.
- Non-standardized Food Products: They do not have specified food standards and their safety parameters are not known. Example: Traditional food.

Food Laws and Regulations

- Food Law is defined as set of standards and guidelines that govern the process of food production, distribution, sales and consumption. Its objective is to protect consumer's health and enhance the production rate.
- Food Regulations contain detailed provisions related to different categories of products defined separately in each set of regulations. It includes minimum quality requirements to ensure that the food is unadulterated and processed in proper hygienic conditions.
- Every country establish own food regulatory framework depending upon the needs of food sector or whether it adopts International standards developed by Codex Alimentarius Commission (CAC) and Food Agriculture Organization (FAO) of the United Nations.

FSSA, 2006

India covers a large geographic area with lots of season that allow a wide variety of flora and fauna to flourish. It also blessed with fresh and marine water sources as well that provides lists of fresh meat produce to enjoy by coastal inhabitant and also serve purpose of business by traders. Not only this, availability of widest milk sources (cow, buffalo, sheep, goat, camel and yak) and spices are the reasons that seeks an umbrella to regulate these commodities under a single agency. Thus rise of FSSA (Food Safety and Standard Act) in 2006 make it convenient by drafting various science based standards and guidelines related to the manufacturing and distribution of food items. This act was introduced in 2006 but came into force in 2011. It protects consumer health and ensured the production of good quality of food. It also regulates the matters related to the import of food, functions allotted to the authorities, penalties and surveillance etc. (*Pardeshi, 2019*).

The following acts were repealed after the commencement of FSS Act, 2006:

- Prevention of Food Adulteration Act, 1954
- Fruit Products Order, 1955
- Meat Food Products Order, 1973
- Vegetable Oil Products (Control) Order, 1947
- Edible Oils Packaging (Regulation) Order 1998
- Solvent Extracted Oil, De-oiled meal and Edible Flour Order, 1967
- Milk and Milk Products Order, 1992

Table 1: Regulatory framework of all food laws

Food Laws	Implementing Ministry	Area of Food
Prevention of Food Adulteration Act, 1954	Ministry of Health and Family Welfare, Directorate General of Health Services	All food commodities
Fruit Products Order, 1955	Ministry of food processing, Govt. of India food and Nutrition Board	All Fruits and Fruit beverages
Meat Food Products Order, 1973	Ministry of Agriculture and Rural Development, Directorate of Marketing and Inspection	All Meat and Meat Products
Vegetable Oil Products (Control) Order, 1947	Ministry of Industry, Govt. of India. Directorate of Vanaspati	Vanaspati and edible oils used for Hydrogenation
Edible Oils Packaging (Regulation) Order 1998	Ministry of Civil Supplies and consumer affairs	All edible oil products
Solvent Extracted Oil, De-oiled meal and Edible Flour Order, 1967	Ministry of Civil Supplies and consumer affairs	All edible oils/Flours and similar products
Milk and Milk Products Order, 1992	Ministry of Agriculture, Govt. of India, Milk and Milk Products Advisory Board	All Milk Products and fluid Milk

Prevention of Food Adulteration (PFA) Act, 1954

Fresh, pure and nutritious food is most important for the health as it prevents malnutrition and reduces the risk of various diseases like diabetes, cancer, stroke etc. But most of food gets adulterated to make more profit and to increase its quantity while reducing its quality by the addition of inferior substances replacing the valuable ones such as addition of brick powder and artificial color (Sudan red) to the red chili powder, Metanil yellow used for coloring dal and turmeric etc. which leads to terrible health problems. To prohibit this kind of anti-social evil Government introduce PFA (Prevention of Food Adulteration) Bill in the Parliament which defines various food standards and guidelines to regulate the manufacture, distribution, storage and import of the food products and make provisions for prevention of food adulteration/contamination, enrichment of flour, bread with vitamins and minerals, addition of vitamin "C" in certain foods, vitaminisation of vansapati, and iodization of salt.

Table 2: Different types of Contaminates/Adulterants

Food Items	Type of Adulterants/Contaminates
Coffee and Tea	Coat tar dyes, husk, tamarik husk, excessive stuff, sand and grit and used tea dust.
Milk	Water and starch etc.
Non-Alcoholic beverages	Non-permitted colors, ducin, saccharin, arsenic, lead and copper.
Starchy Foods	Foreign starches in arrowroot, sand, dirt, etc.
Baking Powder	Citric acid
Spices	Coal tar dyes, sand, grit, lead chromate in haldi, and excessive stalky and woody matter in zeera.
Mustard seeds	Argemone seeds which can cause epidemic dropsy.
Dals	Kesari dal which can cause lathyrism coal tar dyes.
Groundnut	Aflatoxin can cause cirrhosis of liver
Vanaspati	Excessive hydration of rancid stuff, animal fat and foreign flavor etc.
Oils	Mineral oil potential carcinogenic, argimone oil.

Approximately 82 State Level Food Laboratories and 4 Central Food Laboratories have been established at Ghaziabad, Pune, Mysore and Calcutta. Central government defines the standards related to the quality, production, distribution, sale and packaging of food item send in disputed cases, the samples were referred to the Central Food Laboratories for the final opinion. Following amendments in PFA act were made by the Central Committee of Food Standards (CCFS):

- Prevention of Food Adulteration (Amendment) Act, 1964
- Prevention of Food Adulteration (Amendment) Act, 1971
- Prevention of Food Adulteration (Amendment) Act, 1976
- Prevention of Food Adulteration (Amendment) Act, 1986

Fruit Products Order (FPO), 1955

Manufacturing of fresh produce on large scale without following proper hygienic and sanitation procedures increases the chances of contamination as it involves number of processing steps (peeling, slicing, chopping, coring or trimming) in which plant fluids are released and it acts as a nutritive medium for the growth of pathogens. Use of contaminated water for the washing or chilling purpose also spread contamination in a large volume and excess amount of preservatives are used to enhance the shelf-life which are harmful for health. To establish standards for fresh produce processors, Essential Commodities Act revised Fruit Product Order in 1955 to set the quality control guidelines which are compulsory for all manufacturers to keep hygienic condition during manufacturing process for the production of safe and good quality of products. The quality guidelines include Personnel hygiene, sanitary conditions of premises, Portability of water, Quality control facility, Product standards and Limits for preservatives, additives etc. The FPO mark is a mandatory mark for all fruit business operators, it includes following fruit products:

- Fruit Juice, Squash, Nector, Fruit Syrup, Pulp Concentrate, Aerated water containing fruit juice etc.
- Barley Waters (Lemon, orange, grape fruits etc)
- Dehydrated fruits and vegetables
- Bottled or canned fruits vegetables
- Tomato Juice and Soups
- Tomato puree, paste and ketchup
- Jams and Fruit Cheese
- Fruit Jellies and Marmalades
- Pickles in vinegar/oil
- Candied and crystallized fruits and peels
- Chutneys etc.

Meat Food Products Order (MFPO), 1973

Meat is a rich source of protein, zinc, iron and vitamin B12 which is widely consumed as a part of balanced diet but due to the emerging risk of communicable diseases that transfers to human through meat, consumers demand products that are processed in sanitary environment without the risk of contamination. For this purpose, Ministry of Food Processing implements Meat Food Products Order (MFPO), 1973 to manage the processing of meat products including fish and poultry products and it aims to keep cleanliness in slaughterhouses, antemortem and postmortem examination, in-process inspection and end product testing. In most of the states, slaughtering of cows is banned and beef export is prohibited.

Vegetable Oil Products Order (Control), 1947

With growing population the consumption rate of vegetable oil is increased rapidly. In 2017 it was reached at 23 million tons and expected to exceed 34 million tons by 2030. Most of oil varieties including soy oil, palm oil and sunflower oil are imported from Malaysia, Indonesia and South America. To fulfill the consumers demand and to enhance oil production rate government replaced the Vegetable Oil Products (Control) Order, 1947 and Vegetable Oil Products (Standards of Quality) Order, 1975 by Vegetable Oil Products (Regulation) Order, 1998. It properly regulates the processing of oil products through Directorate of Vanaspati, Vegetable oils and Fats, Department of Food and Public Distribution.

Edible Oil Packaging (Regulation) Order 1998

For oil packaging different types of packaging material is used to extend its shelf life. It provides moisture and oxygen barrier and prevents rancidity. Specified minimum requirements were formed related the use of packaging material to ensure consumer protection. To maintain the quality of packaged oil Edible Oils Packaging (Regulation) Order has been introduced by government in 1998 under Essential commodities act, 1955. It makes the edibles oils packing at predetermined prices and sold in retail.

The Solvent Extracted Oil, De oiled Meal, and Edible Flour (Control) Order, 1967

To satisfy the demand of growing population large amount of oil cakes (castor, peanut, sunflower, cottonseed, linseed) are processed. Extraction of oil from oil-bearing substances includes the use of solvents like hexane to achieve highest oil yield. So, Solvent Extracted Oil, De oiled Meal, and Edible Flour (Control) Order, 1967 set quality standards to provide consumer protection by inhibit the use of oil without get refined. It also defines standards for the use of solvent (hexane) during the oil extraction process and regulates the manufacturing and transportation of solvent extracted oils etc.

Milk and Milk Products Order (MMPO), 1992

After being collected from the dairy farms milk is processed within the few hours under the hygienic conditions to maintain its quality. So, various quality control parameters were established by department of Animal Husbandry and Dairying (GOI) under MMPO, 1992 sec. 3 of the Essential commodities act, 1955 to provide safest quality products. Its aim is to enhance the quality of milk by maintaining hygienic conditions throughout the processing. Various changes were done in milk and milk products order, 1992 to increase the growth rate of dairy sector faster.

FSSR, 2011

Food Safety and Standards Regulations, 2011 comprises of all major guidelines and regulations notified by the Government of India and came into force on 5th August, 2011. The all regulations are further reviewed and suitable changes were done according to the latest developments in the field of food science, formulation of new products and additives, advancement in processing methods and food consumption patterns.

Table 3: FSSR, 2011

S. No.	Ref. Section	Regulations notified	Area
1.	92 (2) (e)	Food Safety and Standards (Food Products Standards and Food Additives) Regulations, 2011 (57 Amendments)	Notify standards and guidelines in relation to articles of food meant for human consumption under sub-section (2) of sec. 16
2.	92 (2) (h)	Food Safety and Standards (Food Products Standards and Food Additives) Regulations, 2011	Limits of additives under section 19
3.	i. 92 (2) (i) ii. 92 (2) (j)	Food Safety and Standards (Contaminants, Toxins and Residues) Regulations, 2011	Limits of quantities of contaminants, toxic substances. Under Section 20
		14 Amendments	Tolerance limit of pesticides, veterinary drug residues. Under Section 21
4.	92 (2) (k)	Food Safety and Standards (Packaging and Labeling) Regulation, 2011 8 Amendments Revised Food Safety Standards (Labeling and Display) Regulations, 2019 has been draft notified.	The manner of marketing and labeling of foods under section 23
5.	i. 92 (2) (q)	Food Safety and Standards (Laboratory and Sampling Analysis) Regulation, 2011	Procedure in getting food analysed, details of fees etc. Under sub-section (1) of section 40
	ii. 92 (2) (r)	1 Amendment	Functions, procedures to be followed by food laboratories under sub-section (3) of section 43
6.	92 (2) (v)	Food Safety and Standards (Prohibition and Restriction for Sale) Regulation, 2011 (5 Amendments)	Any other matter which is required to be, or may be, specified by regulations or in respect of which provision is to be made by regulations.

Table 4: New FSSR notified

S. No.	Ref. Section	Regulations notified	Area
1.	22 (1)	Food Safety and Standards (Health Supplements, Nutraceuticals, Food for Special Dietary Use, Food for Special Medical Purpose, Functional Food and Novel Food) Regulation, 2011	Foods for special dietary uses of functional foods or nutraceuticals or health supplements
2.	92 (2) (m)	Food Safety and Standards (Food Recall Procedure) Regulations, 2017	Conditions and guidelines related to food recall procedures under subsection (4) of section 28
3.	16 (2) (d)	Food Safety and Standards (Import) Regulations, 2017 (1 Amendment)	The procedure and the enforcement of quality control in relation to any article of food imported into India
4.	22 (4)	Food Safety and Standards (Approval for non-specified food and food ingredients) Regulations, 2017	Proprietary and Novel food
5.	22 (3)	Food Safety and Standards (Organic Food) Regulations, 2017	Organic Food

FAO

To lead healthy lifestyle, nutrition plays an important role. Production of complementary foods provides positive impact on health and reduces the risk of malnutrition. FAO (Food and Agriculture Organization) arrange various workshops and programs to aware consumers related to nutrients requirements and labeling on the products. It is the oldest agencies of the United Nations (UN), established in 1945. FAO plays major role in the development of food and agriculture including forestry and fisheries practices. It makes continuous efforts to defeat hunger, improve nutrition, to achieve food security and helps to reduce rural poverty by enhancing the production rate of all agriculture products. It is an inter-governmental organization (IGO) which provides technical assistance, arranges educational programs, maintains information and statistics related to the production, consumption and trade of the agricultural commodities and also publish a number of yearbooks and research bulletins.

Table 5: Departments of FAO and their role

S. No.	Departments of FAO	Role
1.	Agriculture and Consumer Protection Department	Promotes agriculture to reduce poverty and ensuring safe food production.
2.	Climate, Biodiversity, Land and Water Department	Promotes sustainable management practices – land, energy, water, biodiversity, soil and genetic resources.
3.	Corporate Services, Human Resources and Finance Department	Provide support to the entire FAO organization.
4.	Economic and Social Development Department	Promotes economic development through the internal production and trade.
5.	Fisheries and Aquaculture Department	Promotes the management of fishing and aquaculture.
6.	Forestry Department	Promotes the management of forestry resources.
7.	Technical Cooperation Department	Provide support to the member countries and answer the food and agriculture related problems.

Core Functions of FAO includes:

1. Collecting, evaluating and disseminating the information to its members related to nutrition, agriculture and food.
2. Developing international norms and standards.
3. Adopting improved agriculture methods which help to enhance the production rate.
4. Provides emergency and post-emergency assistance through the global network of experts.

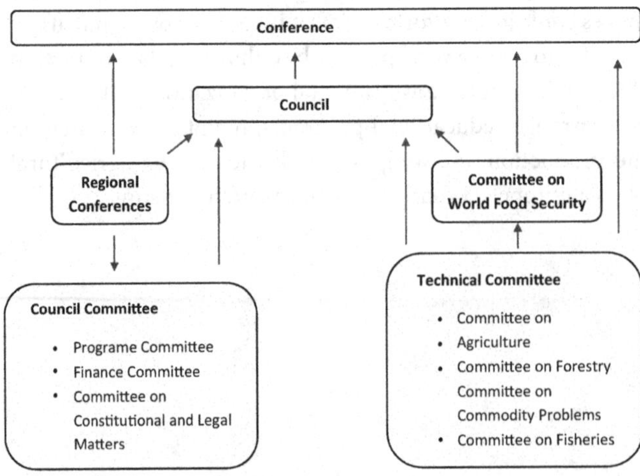

FAO has 194 member states and its headquarters is located at Rome, Italy. It consists of three governing bodies: Conference, Council and Director-General. Jose Graziano da Silva is the current Director General of FAO. It further consists of Council committee, Technical committee and regional conferences.

- Conference plays a supreme role in policy making, approves work programs and also appoints Director-General.
- The Council acts as an executive organ on the behalf of Conference and also able to make decisions. Council committees assist and advice the council on financial, operational and legal administrative matters.
- Technical committees provide proposals and information to the council and conference on Agriculture (COAG), Committee on Forestry (COFO), Committee on Commodity Problems (CCP) and Committee on Fisheries (COFI).
- Five Regional Conferences were established at Europe, Asia and the Pacific, Latin America and the Caribbean, Europe, Africa and the Near East.

ISO

Standardization plays an important role to make continuous improvement in every aspect of life. It helps to provide safe and good quality of products to the consumers and also bring economic benefits. Various international standards were established by International Organization for Standardization (ISO) related to food quality and environment protection. It was established in 1947 as a voluntary and independent international organization, located in Switzerland. It comprises of 164 national standards bodies that form a network of standards institutes and it published 2281 International Standards related to safety and food quality, agriculture, and healthcare technology (Heires, 2008). It also helps to facilitates trade practices and it has 3 membership categories:

- **Full Members:** Full members participate in technical and policy meetings of ISO and also have the right of voting. They can sell and adopt ISO standards nationally. ISO has 120 Full members or Member bodies including India, Italy and Australia.
- **Correspondent Members:** Correspondent members attend the technical and policy meetings of ISO as observers and they can adopt or sell the ISO international standards nationally. ISO has currently 39 correspondent members.
- **Subscriber Members:** Subscriber members cannot participate in the ISO work and do not adopt or sell ISO standards nationally. ISO has three subscriber members.

ISO Standards

The main ISO standards are ISO 9000 (Quality Management), ISO 14000 (Environmental Management), ISO 22000 (Food Safety Management).

- ISO 9000: It formulates a set of international standards related to quality management and quality assurance for improving the product quality and enhancing the customer satisfaction (*Trienekens and Zuurbeir 2008*).

The ISO 9000 family contains these standards:

- ISO 9001:2015: Quality Management Systems—Requirements
- ISO 9000:2015: Quality Management Systems—Fundamentals and Vocabulary (definitions)
- ISO 9004:2009: Quality Management Systems—Managing for the Sustained Success of an Organization (continuous improvement)
- ISO 19011:2011: Guidelines for Auditing Management Systems
- ISO 14000: It refers to a family of international environmental management standards and guidelines and establishes various environmental management policies. It handles the environment related issues and based on a Plan-Check-Do-Review-Improve cycle.

Family of ISO 14000 Standards:

- ISO 14004 – General guidelines on principles, systems and support techniques
- ISO 14006 – Guidelines for incorporating eco-design
- ISO 14015 – Environmental assessment of sites and organizations (EASO)
- ISO 14020 – Environmental labels and declarations
- ISO 14031 – Environmental performance evaluation
- ISO 14040 – Life cycle assessment
- ISO 14050 – Vocabulary
- ISO 14063 – Environmental communication
- ISO 14064 – Greenhouse gases
- ISO 19011 – Guidelines for auditing management systems
- **ISO 22000:** It defines as a Food Safety Management System (FSMS) that maintains food safety throughout the food chain, from farm to fork. It incorporates Good Manufacturing Practices (GMPs), ISO 9001:2000 elements and Hazard Analysis Critical Control Point (HACCP) principles. Its objective is to eliminate all type of food hazards and ensures the production of safe food. It covers all organizations that produce, process or manufacture food or feed, such as: Agricultural producers, Processors, Feed and food manufacturers, Retailers, Food outlets and caterers, Equipment manufacturers, Transportation operators,

Biochemical manufacturers, Storage providers and Packaging material manufacturers etc. (*Rees and Waston 2000*).

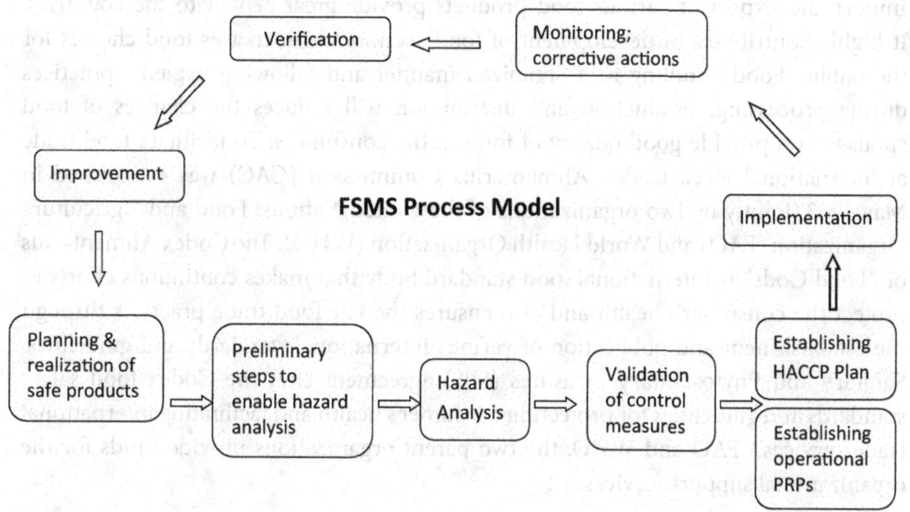

ISO 2000 Standards

- ISO 22000:2005 FSMS – Requirements for any organization in food chain
- ISO/TS 22002 – 1:2009 PRs on food safety – Food manufacturing
- ISO/TS 22003:2007 FSMS – Requirements for bodies providing audit and certification of FSMS.
- ISO/TS 22004:2005 FSMS – Guidance on the application of ISO 22000:2005
- ISO 22005:2007 Traceability in the feed and food chain – General principles and basic requirements for system design and implementation.

JECFA

Joint FAO/WHO Expert Committee (JEFCA) is an international expert committee established in 1956. It evaluates the proportion of food additives, naturally occurring toxicant, contaminants and residues of veterinary drugs in food. Based on JEFCA evaluation, Codex Committee on Food Additives and Contaminants (CCFAC) develops food chemical standards and establishes acceptable limits - Acceptable Daily Intakes (ADIs) for substances like food additives etc. On the basis of toxicological evaluation, it also defines tolerable limits – Provisional Tolerable Weekly Intakes (PTWI) or Provisional Maximum Tolerable Daily Intakes (PMTDI). JEFCA has evaluated 1300 food additives, 25 contaminants and toxicants and approximately 80 residues of veterinary drugs till May 2004. It helps other countries by providing a expert advice for the formulation of their own food safety measures.

Codex alimentarius commission

Import and export of various food products provide great benefit to the countries. It highly contributes in development of food section and increases food choices for the public. Food handling in a organized manner and following hygienic practices during processing, production and distribution will reduces the chances of food spoilage and provide good quality of food to the consumers. To facilitate food trade at international level, Codex Alimentarius Commission (CAC) was established in May 1963, jointly by two organizations of the United Nations: Food and Agriculture Organization (FAO) and World Health Organization (WHO). The Codex Alimentarius or "Food Code" is international food standard body that makes continuous efforts to protect the consumer's health and also ensures the fair food trade practices through the establishment and publication of various international standards and guidelines. Sanitary and Phytosanitary Measures (SPS) agreement cites the Codex food safety standards and guidelines for protecting consumer's health and facilitating international trade services. FAO and WHO, the two parent organizations provide funds for the organizational support services.

Organizational structure

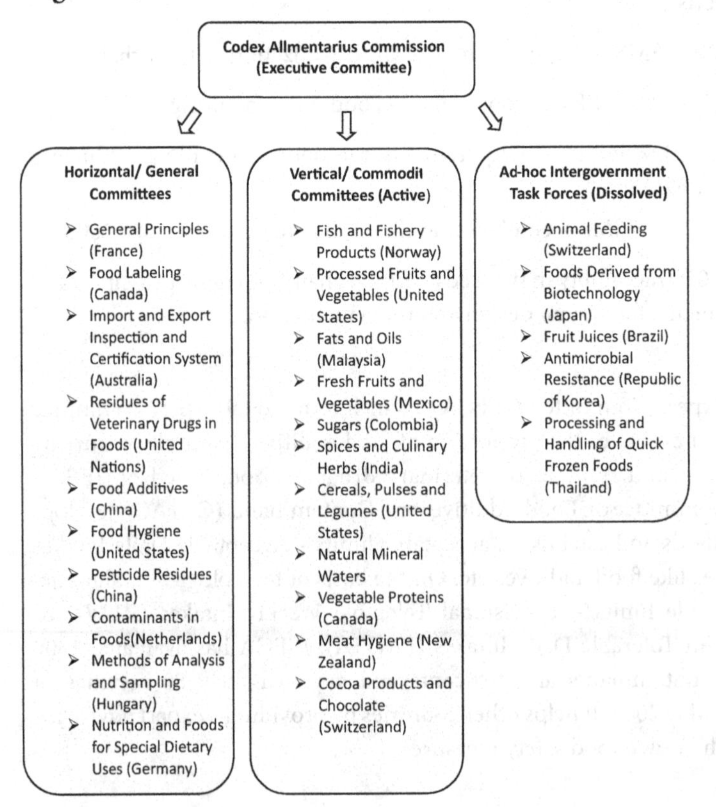

Codex Allmentarius Commission
(Executive Committee)
Horizontal/ General Committees
General Principles (France)
Food Labeling (Canada)
Import and Export Inspection and Certification System (Australia)
Residues of Veterinary Drugs in Foods (United Nations)
Food Additives (China)
Food Hygiene (United States)
Pesticide Residues (China)
Contaminants in Foods(Netherlands)
Methods of Analysis and Sampling (Hungary)
Nutrition and Foods for Special Dietary Uses (Germany)
Vertical/ Commodit Committees (Active)
Fish and Fishery Products (Norway)
Processed Fruits and Vegetables (United States)
Fats and Oils (Malaysia)
Fresh Fruits and Vegetables (Mexico)
Sugars (Colombia)
Spices and Culinary Herbs (India)
Cereals, Pulses and Legumes (United States)
Natural Mineral Waters
Vegetable Proteins (Canada)
Meat Hygiene (New Zealand)
Cocoa Products and Chocolate (Switzerland)
Ad-hoc Intergovernment Task Forces (Dissolved)
Animal Feeding (Switzerland)
Foods Derived from Biotechnology (Japan)
Fruit Juices (Brazil)
Antimicrobial Resistance (Republic of Korea)
Processing and Handling of Quick Frozen Foods (Thailand)

FAO/WHO Regional Coordinating Committees

- Africa (Kenya)
- Asia (India)
- Euroe (Netherlands)
- North America and South West Pacific (Vanuatu)
- Latin America and the Caribbean (Chile)
- Near East (Iran)

Organizational structure

Currently CAC has 189 Codex members consists of 188 member countries and The European Union. In 1964, India became the member of CAC. The development work of the commission is assisted by the executive committee and further, a set of committees were established to perform the whole work in a systematic manner. There are permanent committees and also some committees were established on ad hoc basis. The Codex Alimentarius made standards for raw, semi-processed and processed foods including food distribution work. Codex includes provisions related to residues of pesticides and veterinary drugs, food hygiene, contaminants, food additives, labeling and presentation, analysis and sampling methods etc.

APEDA

India is world's largest agricultural exporter as its diverse climate offers the production of different varieties of agricultural produce including fresh fruits and vegetables. Integration of surplus agriculture products to the global chain not only allows farmers to get better income but also helps the government to earn foreign exchange. For expanding agricultural export, APEDA (Agricultural and Processed Food Product Export Development Authority) was established in 1985 under the Agricultural and Processed Food Product Export Development Authority Act. The Act (2 of 1986) came into effect from 13th February, 1986 by a notification issued in the Gazette of India: Extraordinary: Part-II [Sec. 3(ii): 13.2.1986). The Authority replaced the Processed Food Export Promotion Council (PFEPC). It is a government organization that provides information and financial assistance for the development and promotion of export of various scheduled products. To promote the export of agro-products "Agriculture Export Promotion Scheme of APEDA" provides guidelines and financial assistance to the exporters under sub-sections of the scheme- Infrastructure Development, Quality Development, Market Development and Transport Assistance. It provides services to the agro export community through its head office located at New Delhi, five Regional Offices- Mumbai Kolkata, Hyderabad, Bangalore & Guwahati and 13 Virtual Offices at Bhubaneswar (Orissa), Thiruvananthapuram (Kerala), Chandigarh, Kohima (Nagaland), Chennai (Tamil Nadu), Srinagar (J&K), Raipur (Chhattisgarh), Bhopal (Madhya Pradesh), Panaji (Goa), Lucknow (Uttar Pradesh), Imphal (Manipur), Ahmedabad (Gujarat) and Agartala (Tripura) (*Dudeja and Singh, 2017*).

According to the Agricultural and Processed Food Product Export Development Authority Act, APEDA was assigned with the following functions:

- Development of industries and promoting the export of scheduled products by providing the financial assistance through the subsidy schemes.
- Registration of exporters of specified products on payment of required fees.
- Inspection of meat products in the processing plants and slaughter houses to ensure the quality of such products.
- Improving the packaging and marketing of scheduled products.
- Providing trainings in industries related the scheduled products.

Products which are specified in the APEDA act are known as scheduled products viz Fruits, Vegetables, Meat and Poultry Products, Bakery Products, Jaggery, Honey and Sugar, Cocoa products, Beverages including both alcoholic and non- alcoholic, Walnuts, Groundnuts and Peanuts, Guar Gum, Cereal products, Floriculture products, Chutneys and Pickles, Medicinal and Herbal plants. In addition, APEDA is also responsible to monitor the import of sugar as well.

BIS

Bureau of Indian Standards (BIS) came into existence on 1 April 1987 through Parliament Act 1986. It is the National Standard Body of India responsible for the preparation and implementation of different standards and certification schemes regarding the quality of different products. It contributes towards the conformity assessment and process standardization for the production of goods of better quality, less health hazards, safe for environment and promotes export and import practices.

Organizational network

It has five Regional Offices (ROs) located at Chandigarh (Northern), Mumbai (Western), Chennai (Southern), Kolkata (Eastern) and Delhi (Central) and main headquarter is located at New Delhi. Under regional offices there are various Branch Offices (BOs). These branch offices are responsible for maintaining the link between the State Governments, technical institutions, companies and consumer organizations etc and they are located at different 28 locations namely Ahmadabad, Bhopal, Bhubaneswar, Bengaluru, Chandigarh, Chennai, Coimbatore, Delhi, Dehradun, Faridabad, Guwahati, Ghaziabad, Himachal Pardesh, Hubli, Hyderabad, Jammu, Jamshedpur, Jaipur, Kolkata, Kochi, Lucknow, Mumbai, Nagpur, Pune, Patna, Rajkot, Raipur and Vishakhapatnam.

The main activities of the BIS are described below:

- Standards Formulations
- International Activities

- Product Certification
- Hallmarking
- Laboratory Services
- Training Services (NITS)
- Consumer Affairs and Publicity

Standards formulations

BIS formulates standards related to the different fields that have been categorized as Chemicals, Civil, Electronics and Information Technology, Electro-technical, Food and Agriculture, Metallurgical Engineering, Mechanical Engineering, Management and Systems, Medical Equipment and Hospital Planning, Petroleum Coal and Related Products, Production and General Engineering, Water Resources, Transport Engineering and Textile. Under each Department, one Division Council was established to maintain the working of Sectional committees which helps the industries to enhance the quality of their products and services.

International activities

International Organization for standardization (ISO) is an independent organization, which develops many international standards related to different aspects of quality etc.

International Electro-technical Commission (IEC) plays major role in preparation and publication of International Standards related to electrical technologies. BIS represent India in IEC.

Product certification

Product Certification scheme of BIS and ISI mark also known as a BIS standard mark on a product indicated the conformity of the Indian Standards. BIS license is compulsory for the manufacturing of specific products viz; Food and related products, Diesel engines, Cement, Medical Equipments, Electrical transformers, Steel products, Automobile accessories and household electrical goods etc.

Types of BIS certification schemes

a. **Foreign Manufactures Certification Scheme (FMCS):** For the foreign manufacturers a separate scheme was designed to certify the goods which are manufactured outside the India.

b. **Registration Scheme for Self Declaration of conformity:** In registration scheme, a declaration was made by the manufacturer that his goods conform to the Indian Standards.

c. **Tatkal Scheme:** This scheme employed for those manufacturers who need ISI mark (BIS standard mark) within 30 days as per government notification, they can obtain the BIS certification on the priority basis.

d. **ECO Mark Scheme:** The ECO mark scheme, specially designed for the eco-friendly products which qualify the requirements that given in the specified Indian Standard.

As per the notification released by the Ministry of Electronics and Information Technology (MeitY) on 3 Oct 2012 BIS registration was mandatory for the Electronics and Information Technology (IT) products.

Hallmarking

In April 2000, BIS initiated the gold hallmarking to give assurance to customer about the purity or fineness of gold.

Laboratory services

Eight Central Laboratories was established by BIS for the testing purpose and they are located at Mohali, Mumbai, Guwahati, Patna, Kolkata, Chennai, Sahibabad and Bangalore. There are also NABL accredited laboratories for the conformity assessment of the products.

National Institute of Training for Standardization (NITS)

Through NITS, BIS provides training to the management personnel from companies, consumer organizations and government bodies etc.

Consumer affairs and publicity

Various awareness programs are organized by BIS through Regional and Branch Offices to promote consciousness among consumers and industries related to the product quality, standardization and certification process etc.

MOFPI

India is at no. 1 in the production of bananas, papayas, mangoes and ranks second in the production of rice and wheat but due to lack adequate infrastructure and policies only 2% of produce is processed. For the production of surplus agricultural products and integration of various schemes which provides more opportunities to enhance the growth of processing section, government of India establish Ministry of Food Processing Industries (MOFPI) in 1988. It is responsible for the formulation of laws; regulations and development of agro-food processing sector by the better utilization of the agriculture produce.

Objectives of Ministry

- Development of new food processing techniques to enhance the production rate and reduce the wastage of food.
- Promote the export of processed products.
- Provide policy support for the growth of food processing sector.

- Increase farmer's income by value addition and better utilization of agriculture produce.
- Support R&D sector, for product development and improve the packaging techniques

To achieve these goals, ministry launched schemes like Agro Processing Cluster, Pradhan Mantri Kisan SAMPADA Yojana and Mega Food Parks.

- **Agro Processing Cluster:** This scheme aims to encourage the entrepreneurs to set up various food processing units based on cluster approach by establishing the links between the producers, processors and markets. Project Execution Agency set the Agro-processing clusters and manages projects including DPR (Detailed project report) formulation, creating infrastructure and arranging finance etc.

Each Agro-processing cluster consists of two basic components:

- Basic Enabling Infrastructure: water supply, roads, power supply, drainage etc.
- Core Infrastructure/ Common facilities: cold storages, tetra pack, ware houses, sorting, grading etc).
- **Pradhan Mantri Kisan SAMPADA Yojana:** is a new Central Center Scheme approved by GOI for development of Agro-processing Clusters and Agro-Marine Processing. This scheme enhances the growth of food processing sector, doubles the farmer's income, create employment opportunities in rural areas, reduce wastages and promote the export of processed foods. Schemes implemented under PM Kisan SAMPADA Yojna are Mega Food Parks, Human Resources and Institution, Infrastructure of Agro-processing Clusters, Expansion of food preservation capacities and Quality Insurance Infrastructure (Jairath and Purohit, 2013).
- **Mega Food Parks:** The Mega Food Park Scheme concentrate on bringing farmers, retailers and processors together, maximizing value addition of agriculture produce, increasing farmer's income. This scheme is based on Cluster approach and setting up of new processing techniques. It consists of well-established supply chain which includes collection centers, primary and central processing centers. 40 Mega Food Parks are funded under this scheme and around 17 Mega Food Parks are operational which includes:

1. Srini Mega Food Park, Chittoor, Andhra Pradesh.
2. Godavari Mega Aqua Park, West Godavari, Andhra Pradesh.
3. North East Mega Food Park, Nalbari, Assam.
4. Gujarat Agro Mega Food Park, Surat, Gujarat.
5. Cremica mega Food Park, Una, Himachal Pradesh.
6. Integrated Mega Food Park, Tumkur, Karnataka.

7. Indus Mega Food Park, Khargoan, Madhya Pradesh.
8. Paithan Mega Food Park, Aurangabad, Maharashtra.
9. Satara Mega Food Park, Satara, Maharashtra.
10. MITS Mega Food Park, Rayagada, Odisha.
11. International Mega Food Park, Fazilka, Punjab.
12. Greentech Mega Food Park, Ajmer, Rajasthan.
13. Patanjali Food and Herbal Park, Haridwar, Uttarakhand.
14. Himalayan Mega Food Park, Udham Singh Nagar, Uttarakhand.
15. Jangipur Bengal Mega Food Park, Murshidabad, West Bengal.
16. Tripura Mega Food Park, West Tripura, Tripura.
17. Smart Agro Mega Food Park, Nizamabad, Telangana.

FSSAI

Any type of food whether it is processed or semi-processed including genetically modified food should follow minimum safety requirements to provide confidence to the consumers that it doesn't cause any harm. It is compulsory for all food business operators to comply with the standards established by Food Safety and Standards Authority of India (FSSAI). It is an autonomous body which regulates the whole food processing chain to ensure the production of safe food. Ministry of Health & Family Welfare (GOI) is the administrative Ministry of FSSAI. It has eight regional offices located in Delhi, Mumbai, Chandigarh, Lucknow, Kolkata, Cochin, Chennai and Guwahati and its main head office is situated at New Delhi. It also has 19 referral laboratories, 88 State/UT laboratories and 172 NABL accredited private labs *(FSSAI, 2011).*

FSSAI Structure

The Chairperson and CEO of FSSAI have been appointed by Government of India. Currently, Ms. Rita Teaotia is the Chairperson and Mr. Pawan Kumar Agarwal is the CEO of FSSAI (*Shukla, et al. 2014*).

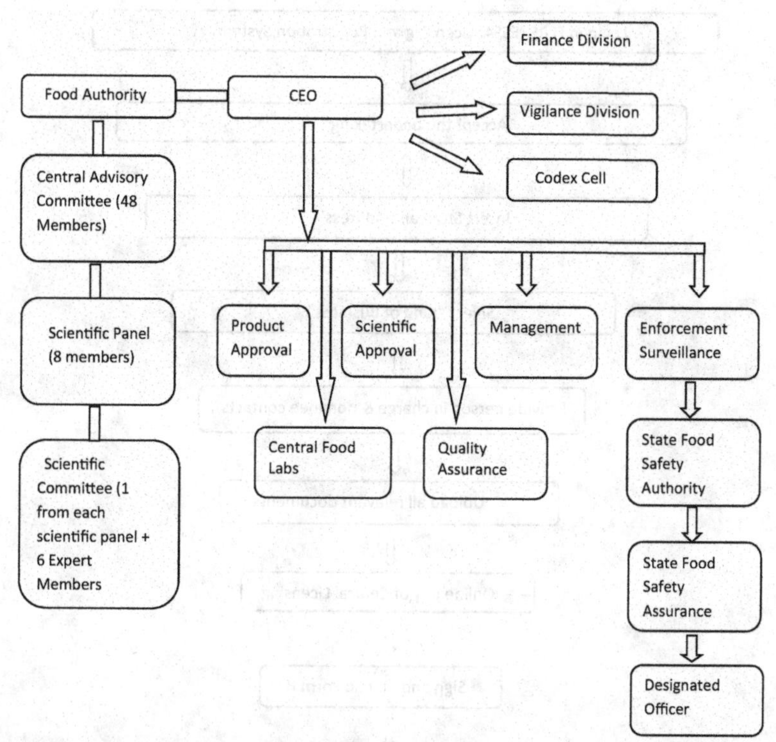

Registration and Licensing: First step to get FSSAI License and FSSL number is registration and it will take 32- 52 working days. Main types of FSSAI License – Basic, State and Central License. The license validity is 1-5 years and its renewal is possible. The basic registration documents will include latest passport sized photographs, identification proof (PAN card), a copy of Property papers and rental agreement.

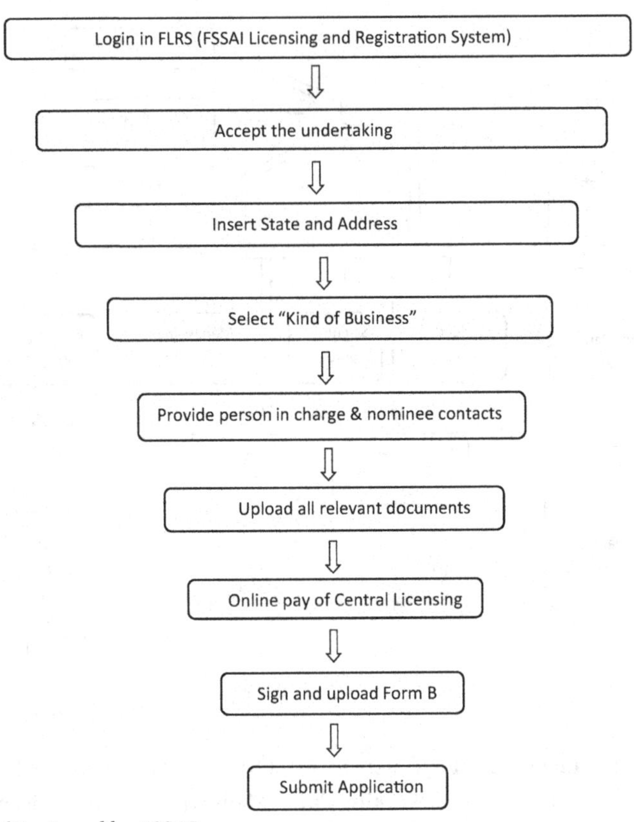

Table 6: Penalties issued by FSSAI

Section	Offence	Fine (Rs.)
50	Food not of quality demanded by purchaser; Not in compliance with the Act	2 Lakhs
51	Sub-standard food	5 Lakhs
52	Misbranded food	3 Lakhs
53	Misleading advertisement	10 Lakhs
54	Food contains extraneous matter	1 Lakh
55	Failure to comply with FSO direction	2 Lakhs
56	Unhygienic Processing or Manufacturing	1 Lakh

References

Chengappa, P.G. (2004). Emerging trends in agro-processing in India. *Indian Journal of Agricultural Economics*, *59*(902-2016-68029).

Dudeja, P., & Singh, A. (2017). Role of government authorities in food safety. In *Food Safety in the 21st Century* (pp. 243-256). Academic Press.

FSSAI (Food Safety and Standards Authority of India). About FSSAI. 2011.

Heires, M. (2008). The International Organization for Standardization (ISO). *New Political Economy*, *13*(3), 357-367.

Jairath, M. S., & Purohit, P. (2013). Food safety regulatory compliance in India: A challenge to enhance agri-businesses. *Indian Journal of Agricultural Economics*, *68*(902-2016-66830), 431-448.

Keservani, R. K., Sharma, A. K., Ahmad, F., & Baig, M. E. (2014). Nutraceutical and functional food regulations in India. In *Nutraceutical and Functional Food Regulations in the United States and Around the World* (pp. 327-342). Academic Press.

Pardeshi, D. S. K. (2019). Food Safety and Standards Act (FSSA) 2006 (34 OF 2006): ITS Legal Provisions, Penalties and Offences. *Journal Homepage: http://www. ijesm. co. in*, *8*(7).

Rees, N., & Watson, D. (2000). *International Standards for Food Safety*. Springer Science & Business Media.

Shukla, S., Shankar, R., & Singh, S. P. (2014). Food safety regulatory model in India. *Food Control*, *37*, 401-413.

Trienekens, J., & Zuurbier, P. (2008). Quality and safety standards in the food industry, developments and challenges. *International Journal of Production Economics*, *113*(1), 107-122.

2

Evolution of Bakery and Confectionary Technology

Anurag Singh

Department of Food Science and Technology, National Institute of Food Technology Entrepreneurship and Management (NIFTEM), Kundli Sonipat, Haryana, India

---

Introduction

Baking is not a new concept to us. Baking process dates thousand years back since the Paleolithic age. During those days, the baking was done by simply grinding the grains, mixed with water and baked on the hot rocks. The bread produced that time was not leavened but flat. Ancient Egyptians have the credit for development of baking between 2600-1200 BC. Roman Empire also had the popularity of the art of baking and bakeries were in existence in Europe too during Middle Age. Baking was considered a highly respected profession under the Roman Empire. The art of baking has travelled a long journey since then. The bakery and confectionary has undergone a gradual evolution with time. Great technical progress stared in the nineteenth century. Development of automatic machines reduced the task of the manual labour and because of this development the bakers could perform many more tasks with the machines. Today's baking has a dependency on the machines. All the work is performed by the automated machines whereas the bakers only control the machine operations. The baking process has become easier, the efficiency has increased and the quality of the products has improved because of these high-tech machines. The preparation of bakery products using traditional methods has also improved over the time to minimize the energy and time consumption. Today's consumer is very much concerned about his/her health and look for the healthy options in baked products. The trend has led to various innovative formulations for the baked products so that the health benefits may also be provided not only the calories. This chapter will discuss various advancements and trends in different types of bakery and confectionary products.

Evolution of bread making

Yeast leavened bread is prepared mixing wheat flour with yeast, water, fat and salt to form a visco-elastic dough. The fully developed dough is allowed to ferment. After fermentation, the dough is divided in small pieces of desired size, rounded like a sphere, moulded like a cylinder and kept in a greased pan. The moulded dough is kept for proofing and baked thereafter. Yeast leavened bread making might have been accidently invented when the wild yeast from the atmosphere could have introduced in a dough left in open for baking. The prepared bread had the texture different from the flat/ un-leavened bread. It may be considered as the start of the yeast leavened bread. Bread making process became a controlled process later with the knowledge about yeast and the process of fermentation. Today's bread making had evolved over the time. There are various additives those can be used in the preparation of breads as well as there are different machineries available for continuous bread making.

Processes for bread preparation can be divided in two main categories: *Long fermentation processes (slow processes)* and *Rapid processes*. Rapid processes have evolved to minimize the production time in bread making by making the process a continuous one. These processes are discussed in detail below:

Bread making with long fermentation

A very long fermentation/ resting time is given to the dough after mixing and before dividing in long fermentation processes. There are two processes in this category namely *Straight dough process* and *sponge and dough process*. All the ingredients are mixed in a single step in straight dough process whereas in case of sponge and dough method, part quantity of the dough is fermented for a long time and remaining ingredients are mixed in this fermented sponge in second stage of mixing. These methods are elaborated in detail as follows:

Straight dough process

Straight dough bread making method is the oldest method. All the ingredients are mixed together to form a homogenous mass and kept for bulk fermentation. The bulk fermentation time depends on various factors viz. flour quality, quantity of yeast added, temperature of the dough and bread variety being produced. Dough development is done manually or using low speed mixers. In general, the strong flour that contains higher protein content requires long bulk fermentation period. The dough is developed due to the enzymatic action during the bulk fermentation.

Sponge and dough process

Sponge and dough process is another process that uses the bulk fermentation. Part quantity of the flour (usually 2/3rd) is mixed with a part of water and yeast to form a loose dough or batter that is called as sponge. The prepared sponge is then allowed to ferment for a longer period upto 5 hrs. This fermented sponge is then mixed with the remaining ingredients to develop a proper dough for bread making. This developed dough is processed after a short floor time period.

Rapid bread making processes

Bread can be prepared rapidly using various combinations of additives (active ingredients) and processing methods. Some improvers to assist dough development and to reduce the fermentation time period are used in rapid processes. Various rapid processes are elaborated as follows:

Activated Dough Development (ADD)

Activated Dough Development was developed during early 1960s in United states of America and gained popularity among the bakeries of the USA and the UK. Mixing time of the flour is greatly reduced in this method using a reducing agent viz. L-cysteine Hydrochloride, proteolytic enzymes and ascorbic acid. Addition of an oxidizing agent is also done at the time of mixing irrespective to that is added at flourmill. Natural softening of the dough is compensated by adding extra water whereas normal proving times is maintained by adding extra yeast (1-2%). The effect of L-cysteine is very fast while mixing and reduces the mixing time required upto 50% or more depending on the amount added. Dough development time can also be reduced using proteolytic enzymes. The enzymes may reduce the total mixing time by 15 to 20 per cent during the period of dough development in the mixer as the exposure time to the protein is short but the activity of enzymes keeps on going even after the dough is mixed. As the ADD process develops the dough chemically, the use of low-speed mixers is sufficient for mixing. A short period of bulk fermentation before dividing the dough may be given to achieve a good quality product. Addition of sponges may be done to change the flavour of the bread. For optimum quality of the products, the final dough temperatures are kept in the range of 25-27°C.

Mechanical dough development

The methods mentioned above can be performed using slow speed conventional mixers but the time consumption is very high for mixing. The time required for mixing can be minimized by mixing the dough at high speed in specially designed high speed mixers. Mechanical dough development methods perform mixing using such high speed mixer. The maturation/ development of the dough in these processes is done by adding dough improvers, extra amount of water and mechanical energy. 'Do-maker' process, one of the mechanical dough development methods, was first successfully developed in the 1950s. A continuous mixer was used in the 'Do-maker' method and the prepared bread having fine and uniform crumb was not that much popular. Later another method called Chorleywood Bread making Process (CBP) was developed that is being widely used in bread manufacturing industry.

Chorleywood Bread Process (CBP)

High speed mixers are used in Chorleywood Bread Process (CBP) for mixing and dough development. With a fixed energy input of 11Wh/kg of dough, the mixers are run for 2 and 5 minute. A combination of fast and slow acting oxidizing agents is used to achieve desired level of softening in the dough viz. potassium bromate and potassium iodate. Fat having high melting point, an emulsifier or a combination of fat and emulsifier is added in the dough. The consistency of the dough is adjusted by adding extra water so that it is comparable with that is obtained from bulk fermentation method. Yeast is added in excess to get the effect of proofing similar to that in bulk fermentation. The bulk fermentation process and CBP differ to each other in the use of high speed mixers for dough development instead of a lengthy resting period. Modification in the protein network of the dough is done in both the processes to improve its ability to retain gas from yeast fermentation. The same is achieved within 05 minutes of mixing in case of CBP.

Latest trends in breads

The bakers are preparing newer types of breads using various new ingredients. The latest trends are discussed as follow:

Sourdough bread

Sourdough can be defined as dough that contains water and metabolically active microorganisms, mainly lactic acid bacteria (LAB) and yeast. It is an intermediate product. The metabolic products of LAB developed during the fermentation process improve the sensorial attributes, technological properties, nutritional value, health aspect and also extend the shelf-life (Salovaara, 1988; Hammes and Ganzle, 1998; Di Cagno et al. 2004; De Angelis et al. 2010). The preparation of sourdough bread can be done as per the traditional practice where the flour is added with water and permitted to stand until wild airborne yeasts settled on it and began to ferment it. This "starter" culture of the yeast can be then used to leaven the bread. A portion of this starter is kept for further addition in another batch of bread. Leavening and production of aroma compounds in sourdough bread is due to the yeast present in sourdough. LAB species are mainly accountable for acidification of the dough as well as they contribute in the formation of flavor and leavening of dough. Sourdough bread has a longer shelf-life because of the production of antifungal (Manini et al. 2016) as well as bacteriocins. The sourdough bread possesses greater sensory quality by means of more elasticity (Clarke et al., 2004), internal humidity (Corsetti et al. 2012) and volatile compound concentration (Makhoul et al., 2015), and higher nutritional value because of reducing anti-nutritional compounds as phytase enzyme (Gänzle, 2014).

Frozen dough bread

Use of frozen dough for bread making is one of the latest trends. In this technology, all the ingredients are added and dough is prepared for bread making. The prepared dough is frozen for later use. The dough undergoes a slow proofing. Whenever required for bread preparation, the dough is taken out from the freezer and used. This reduces the time requirement for the mixing and proofing as compared to the traditional bread preparation methods. The retail bakers produce various baked products such as fresh bread, rolls and Danish pastries by using frozen dough more conveniently and economically than the traditional practices of 'scratch baking' that involves weighing of ingredients, mixing, shaping, fermenting and then baking (Stauffer, 1993). The use of frozen dough for preparing various baked products has increased tremendously in last few years. Technology of frozen bakery has continuously developed and the products those were with low volume, coarse texture and shorter shelf life earlier, now a days are safe for consumption up to six months (Rosell and Gómez, 2007).

Partially baked bread

This type of bread is prepared by baking the bread partially till the structure if fixed. Other terms given for this type of bread are part-baked, par baked bread or pre-baked bread. The crumb structure is available in partially baked bread but not a developed crunchy crust. The part-baked bread is required to be stored at low temperature. Partially baked bread requires a very short time for baking. As the part baked product has already gained the shape and size, it requires lesser time to complete the baking. This is why the market of the part baked bread is increasing tremendously. The pre-baking time is usually two thirds of the time required for complete baking. As the part-baked bread is not having any crust and has higher moisture content, it is susceptible to microbial growth and possesses a very short shelf life. Modified atmospheric composition containing 40% CO_2 and 60% N_2 has been tried by Leuschner et al. (1999) to extend the shelf life of the part-baked bread. Polyethylene-polyamide-polyethylene vinyl alcohol bags containing 70% CO_2 were used by Doulia et al. (2000) for packaging of partially baked bread. Partially baked bread exhibited the extended shelf life upto ten days when stored at low temperatures (2-6°C) by Bárcenas and Rosell (2006, 2007). As per some researches the partial baking process developed the bread which had the similar organoleptic and textural properties as that of the bread obtained by a conventional method. However the crumb hardness increased with frozen storage period of pre-baked bread.

Utilization of composite flour for bread making

Wheat is a good source of calories but is lacks in essential amino acids such as lysine and threonine (Dhingra and Jood, 2001). Supplementation of wheat flour with other cereals or pulses can improve the overall nutritional value of the baked products (Sharma et al., 1999). This hypothesis has evolved the use of composite flour for bread preparation. Utilization of composite flour (a blend of wheat flour with corn, barley, cassava and chickpea) has been prominently studied for preparation of composite flour breads (Defloor et al., 1993; Khalil et al., 2000). Flours of Legumes when used in baked products can produce a product with enriched protein with improved amino acid balance (Bojňanská et al., 2012; Mohammed et al., 2012).

Among the legumes many products have been tested to produce the composite flour baked products viz. soy bean flour (Ribotta et al., 2005; Dhingra and Jood, 2002; Basman et al., 2003), chickpea flour (Gômez et al., 2008), germinated chickpea flour (Luz Fernandez and Berry, 1989), germinated pea flour (Sadowska et al., 2003) and lupin flour (Doxastakis et al., 2002; Pollard et al., 2002). Bread prepared using both barley and defatted soy flours, up to 15% level, was also considered as most acceptable, organoleptically and nutritionally (Dhingra and Jood, 2001). Veluppillai et al. (2010) developed a bread using malted rice flour at 35% and wheat flour at 65% that was having a better consumer acceptability and nutritional values. Maize flour at a level of up to 40% and defatted maize germ flour at a level of up to 15% were also explored for bread preparation by Păucean and Man (2013) and the bread obtained was having reasonable acceptance offering a promising, nutritious and healthy alternative to consumers. Addition of 5% pumpkin flour developed bread with high loaf volume having good overall acceptability as mentioned by See et al. (2007). Fruit flours have also been explored for development of composite bread viz. mango seed kernel flour (Menon et al., 2014) and toasted African bread fruit seed flour (Akubor and Obiegbuna, 2014).

Gluten free bread

Wheat contains specific gluten proteins namely glutenin and gliadin. These proteins make wheat suitable for preparation of many products. In the presence of water and mechanical work, these proteins form a continuous phase named *gluten network.* This network not only reduces the stickiness but also provides the extensibility and cohesiveness to the dough. With increasing occurrences of Celiac disease caused by gluten intolerance, a number of gluten free products have been developed. Many alternative cereals grains/ pseudo-cereals have been tried by the researchers such as buckwheat (Wronkowska et al., 2013), rice flour (Peressini and sensidonni, 2011), amaranth (Alvarez-Jubete et al. 2010), quinoa, soy flour (Shin et al. 2013), chick pea flour (Minarro et al. 2013), chestnut flour (Demirkesen et al. 2010), corn flour etc. for bread making. Proteins present in non-wheat flours do not form the network like gluten proteins, therefore some hydrocolloids such as glycerol monostearate, hydroxyl propyl methyl cellulose (HPMC), xanthan gum (XG), carboxy methyl cellulose (CMC) etc. should be added in these flours. These flours have been added with various other proteins such as dairy proteins (whey proteins and caseins) (Krupa-Kozak et al. 2013), egg proteins, soy proteins (Crockett at al. 2011), collagen or lupine (Ziobro et al. 2013), pea (Marco and Rosell, 2008), yeast or even structured corn proteins to increase the protein content. Gluten free bread was developed by Lopez et al. (2004) by using a composite blend of rice flour with corn and cassava starches. The bread developed had a well-structured crumb and pleasant flavour and appearance. Various tuber crops such as cocoyam, cassava, taro and other tuber crops have also been used as alternative raw material for gluten free bread preparation (Giami et al., 2004).

Evolution of cake making

The word cake is of Viking origin, derived from the Norse word "kaka." The baking skills were first exhibited by the ancient Egyptians. In ancient times, the cakes were more or less bread like in appearance and honey was used for sweetening them. Early form of cheesecake was exhibited by the Greeks whereas the Romans developed fruitcakes using raisins, nuts and other fruits. Modern cakes with round shape and having icing on the top were developed in Europe in 17th century. The first icing developed was a boiled mixture of sugar, egg whites, and some flavorings. Cakes became more popular in the 19th century. The cakes were baked using extra refined wheat flour and baking powder instead of yeast during this time. Traditional boiled icings were replaced by butter-cream frostings later on.

The cake can be defined as a product prepared by baking a mixture of flour, butter, egg and sugar etc; and taken as a snack. Cakes are very spongy delicious food product with desired organoleptic characteristics (Singh and Chaudhary, 2016). In its oldest forms, cakes were modifications of breads, but cakes now cover a wide range of products that may be simple or elaborate, and that share features with products like pastries, meringues, custards, and pies. Filling of fruit preserves, nuts or dessert sauces (like pastry cream), iced with butter cream or other icings, and decorated with marzipan, piped borders, or candied fruit can also be done in cakes.

Cake preparation involves baking of a batter prepared by mixing the ingredients together. The volume of batter expands during baking as the volume of the air bubbles contained in batter expand at high temperature as well as due to the effect of leavening agents. The volume expansion caeses when the starch is gelatinized and provides more rigid structure that is difficult to expand. The batter prepared is a type of emulsion of air in a mixture of ingredients. The air incorporation, its distribution and the viscosity of the continuous phase surrounding the air bubbles are the important aspects in cake making. Smaller the gas bubbles, the higher will be their stability in the batter, and the higher will be the final volume of the cake. The viscosity of the mixture should prevent the coalescence of air bubbles as well as should allow the volume rise during baking.

The cakes can be classified in two groups. The first group of cakes has formulations without additional fat and usually no leavening agent. The cake is leavened by incorporating a large number of tiny gas bubbles. Egg albumin plays the major role in these cakes for leavening due to its foam formation capacity. E.g. *sponge cakes* or *chiffon cakes*. The second group of cakes uses oil or fat in the formulation. The use of a leavening agent in such cakes to enhance the volume rise during baking is very common. e.g. *layer* or *pound cakes*

Latest trends in cake making

With the advancement of knowledge about the adverse effects of fat, sugar and refined wheat flour alongwith the increasing demand of the healthy baked products, the use of non-conventional ingredients for cake preparation is also being done. Many researchers have explored the potentiality of various ingredients in cake making.

Use of composite flours for cake making

Ben Jeddou et al. (2017) developed a dough rich in protein and fibre as well as the cake adding potato peel powders. There was a significant improvement in the texture profile of the prepared dough and the cakes, when a blend of potato peel flours was added in wheat flours. The significant reduction in the hardness of the cake was also observed. A blend of 5% potato peel powder with wheat flour gave the best quality cakes and dough. Mango pulp and mango peel powder were mixed with wheat flour at different concentrations by Noor Aziah et al. (2011). The cake prepared by this composite flour had high dietary fiber with low fat, calorie, hydrolysis and predicted glycemic index. On the basis of the sensory quality, sponge cake prepared with 10% mango pulp and 10% mango peel flour was found the best and endorsed the use of these ingredients as fiber-rich ingredients in the preparation of cake and other bakery products with improved nutritional qualities.

Kaur and Sharma (2018) supplemented pumpkin seeds in cake preparation to enhance nutritional content. Pumpkin seed flour was prepared from the raw and the roasted seeds and supplemented the cakes. Cake prepared by supplementation of 20% seed flour was highly accepted after sensory analysis. The cake added with roasted pumpkin seed flour had the maximum protein content i.e. 8.45%. Fiber and ash content (1.80 and 1.59%, respectively) was found higher in the cake supplemented with raw pumpkin seed flour. Similarly iron and zinc content was also higher (2.04 and 0.64mg/100g) in cake supplemented with raw pumpkin seed flour. Various value added cakes and cupcakes were developed by Seth and Kochhar (2018) incorporating partially defatted peanut flour at different levels (10-20%). The developed products were evaluated for sensory attributes by using nine point hedonic rating scale and nutritional composition of the products was assessed by using standard methods. The result showed that plain cake and eggless cakes prepared with 10% addition of partially defatted peanut flour were acceptable whereas cupcakes were acceptable at 15 percent and Multigrain cake at 20 percent level. The nutritional contents were found significantly higher in the developed products.

Tharshini et al. (2018) studied the change in organoleptic and chemical characteristics of cakes prepared with a blend of wheat flour, soybean flour and pomegranate peel powder in various ratios. The cakes prepared by blends of wheat flour with soybean flour and pomegranate peel powder at 85:10:5, 82.5:10:7.5 and 80:10:10 levels had the overall acceptability in the category 'liked very much'. The protein and fat contents of Type I cake were 11.23 and 23.72 per cent, respectively which were significantly higher than that of control cake. Dietary fibres content alongwith calcium, phosphorus, iron, zinc and magnesium content of cake samples made by supplementation was higher than that of control sample. Bajpai et al. (2018) optimized oat cake rusk formulation by varying the oat percentage from 0 to 20 baking at temperature from 170 °C to 190 °C and time from 30 min to 50 min. Cake rusk showed the significantly increased fibre content upto 2.14g/100g. The cake rusk containing 10% oat flour baked at 180 °C for 30 min was found to be of good technological quality and with a good level of acceptance. The nutritional content of the oat cake rusk like Protein, ash and fat were significantly increased than the control sample. Wheat flour was partially replaced by flaxseed flour at different concentrations (5, 15, 30 and 45%) by Moraes et al. (2010) for cake preparation. The evaluation was conducted for the nutritional value, the oxidative stability, and consumer acceptance of cakes. Flaxseed flour upto 30% can be added with wheat flour for cake preparation as per the results of sensory analysis. The cakes so prepared have higher level of functional ingredients viz. dietary fibres and linoleic acid.

Low sugar/ sugar free cakes

The calorific value of the cake is too high because of the use of large quantity of sucrose. That can be lowered by using alternative sweetener for replacing the sugar, fully or partly. Baeva et al. (2000) used microencapsulated aspartame to eliminate sucrose completely and also added bulking agents namely sorbitol, wheat starch and wheat germ for cake preparation. Various characteristics of the cake viz. porosity, springiness, volume and shrinkage of sponge showed a dependency on the amount of the ingredients added. The sponge cake developed with wheat germ exhibited similarity to the control sample with least physical and sensory deviations. The energy value of the diabetic sponge cakes was significantly reduced upto 29% as compared to the ordinary sponge cake. In a study Psimouli & Oreopoulou (2012) prepared different cake formulations using various sugar alcohols viz. mannitol, maltitol, sorbitol, lactitol as well as fructose, oligofructose and polydextrose to replace sugar. Every substitute replaced sugar by an equal amount in cake recipe. The cake batters prepared from these formulations were evaluated for the rheological behavior as well as the comparison was made with respect to physical characteristics of the cakes made thereof with the control cake sample for respective attribute. It was observed that the use of oligofructose, lactitol or maltitol as sugar replacers exhibited similar behavior as of sucrose for batter rheology and increased starch gelatinization temperature. Instrumental measurement and sensory evaluation demonstrated poor quality characteristics in cakes prepared using fructose and mannitol.

In a similar study Miller et al. (2017) studied for the reduction of the calorie content of the baked products utilizing several commercially available alternative sweeteners. They mentioned that sugar alcohols and natural sweeteners can replace sucrose directly as they have similar bulk as sucrose. When high intensity sweeteners are used, a suitable bulking agent is to be supplemented as the sweeteners are light weighted. Cake samples made with maltitol were acceptable whereas erythritol and fructose were not found suitable for cake making. Maltodextrin and polydextrose produced the cakes similar to control one when used as bulking agents. Stevia was not found acceptable due to a disagreeable metallic aftertaste whereas sucralose was acceptable. When sucralose was used in combination with maltodextrin, the cakes prepared were preferred whereas sucralose plus polydextrose was not accepted. The cakes produced using maltitol was liked by consumers w.r.t. flavor, texture and overall liking of cakes similar to sucrose. It was observed that 16% caloric content may be reduced by replacing the sucrose with maltitol in layer cakes.

Egg less cakes

Due to some religious, cultural and health constraints, eggless cake is also in practice. Replacement of egg in cake making was done by Puranik and Gupta (2017) with whey protein concentrate (WPC). Cakes were prepared using WPC 60 (protein content 62.5%) for egg substitution and the effect on physical and sensory attributes were studied. It was observed that the cake so prepared had highest porosity and specific volume as compared to control cake containing eggs. Addition of 7% WPC can prepare the cake same as we get from the egg addition.

Low fat cakes

High amount of fat in cakes make them unhealthy. Several studies have been done to reduce the fat content using some fat replacers in the formulation. Fernandes and Salas-Mellado (2017) used chia mucilage dried at 50°C or lyophilized as fat substitute for developing bread and chocolate cake. The fat replacement upto 50% could be done using chia mucilage for bread and chocolate cake preparation without affecting the technological and physical characteristics. The products were found acceptable with respect to their sensory acceptability. Similarly maltodextrin, inulin, oligofructose, citrus pectin, and microparticulated protein were used for fat replacement in cakes by Psimouli and Oreopoulou (2013). There was no significant difference observed in cake after replacing fat by 35%. When the replacement was increased over 65%, the significant increase of hardness, elasticity, and decrease of volume development was observed. The samples were not acceptable for the taste and flavor when replacement was increased. In a similar study Diez-Sánchez et al. (2018) replaced fat by 30% using soluble, insoluble fiber, or a mix of both fibers. The cake quality, structure, acceptability, and starch digestibility were assessed. Less effect was observed on the structure and quality of the cakes by the addition of soluble fiber (maltodextrin) than the insoluble fiber (potato fiber). There was no significant difference observed in the cakes prepared by the soluble fibres as compared to the control cake. In another study, inulin was used as fat replacer at different concentrations in sponge cake to study the effect on microstructure and physicochemical properties by Rodríguez-García et al. (2012) and upto 70% reduction of fat content was achieved in this study. Andrade et al. (2018) extracted galactomannan from Cassia grandis seeds and used as fat replacer in sponge cake formulations. 75% of the fat in the formulation was successfully replaced using galactomannan @ 1.0% that too without any change in surface microstructure, sensory quality and rheological behavior of the sponge.

Gluten free cakes

The latest trend in cake segment is gluten free cakes. Gluten does not play a key role in preparation of cakes and mostly there in no gluten development. The mixing performed during batter making does not apply enough mechanical work that may result in gluten development. Thus making gluten free cakes is rather simple as compared to bread. It is only replacing the wheat flour with other non-gluten flours e.g. rice or corn for starch.

The main function of the protein present in wheat flour is to produce the dark colour due to maillard reaction during the cake making. Hence, the replacement of wheat flour with the starch sources will adversely affect the colour of the product. The viscosity of batter will differ and air bubble holding capacity will also alter. Protein absorbs more water as compared to starch and if the formulation has only the starch, the batter prepared will be wetter. Therefore the cake formulation having wheat flour replacement with starch, should have some other type of proteins supplemented. Animal protein or the vegetable proteins can be considered for addition. The replacement of wheat flour with rice flour is very common in western countries however the choice of consumer may vary from country to country as the flavor of the gluten free product as compared to the wheat based product is different. Corn and sorghum are also used as substitute of the wheat flour. The particle size of the alternate flour (gluten free) for cake preparation is also very important. Generally the particle size should be smaller than 200 μm but for corn and rice flour it is larger. Use of hydrocolloids can be done to maintain the stability of emulsion in batter of gluten free cakes.

Evolution in biscuit making

Biscuit is one of the most consumed baked products. The major ingredients used for biscuit making are mainly flour, fat and sugar whereas minor ingredients are baking powder, skimmed milk, emulsifier, and sodium meta bisulphite. Biscuits are low moisture product (usually <4%) with a longer shelf life upto six months or longer. There are 10 categories of biscuits as per Manley (2000) *viz.* 1. Bread, pizza and crispbread, 2. Water biscuits and soda crackers, 3. Cream crackers, 4. Cabin biscuits, 5. Savoury crackers, 6. Semi-sweet/hard-sweet, 7. Continental semi-sweet, 8. Short dough (moulded), 9. Wire cut types and 10. Short dough (sheeted).

Broadly the biscuits are classified in two categories namely *Short dough biscuits* and *hard dough biscuits*. The short dough biscuits have higher quantity of fat and sugar as compared to the hard dough biscuits. The hard dough biscuit dough is extensible in nature and can be sheeted without breakage (Manley, 2000) whereas short dough has the shorter texture and difficult to be sheeted. Biscuit preparation from short dough is preferably done using rotary moulding, wire cutting, deposition methods.

Latest trends in biscuit making

With the advancement in use of machines for biscuit manufacturing and the increased awareness of the consumer for the health, various developments have occurred in biscuit making as well. The current trends in biscuit segment can be discussed as bellow:

Gluten free biscuits

Increasing cases of gluten intolerance and occurrence of celiac disease has motivated scientists to develop gluten free biscuits. Sakr and Hussien (2017) developed gluten-free biscuits using a mixture of broken rice flour, sweet chickpeas and date powder in different proportions. The study concluded that date powder can be incorporated in the biscuits up to 20% without affecting the sensory quality. The Supplementation of biscuits with date powder and chickpea flour (20% each) is helpful to cover up to 31.6% of protein requirement, 17.2% of iron requirement and 13.02% of calcium. Man et al. (2014) prepared gluten-free biscuits from mixtures of maize flour (MF), rice flour (RF) and soybeans flour (SF). The blend with flour levels 30:30:40 (MF:RF:SF) led to the highest acceptability. Shahin and Sakr (2016) prepared biscuits from amaranth (flour, wholemeal and hull) with defatted soybean, sorghum and rice. The sensory analysis of the biscuits prepared indicated that no significant difference was found between control sample made from wheat flour. Dates are rich in dietary fibre content. Mishra et al. (2015) developed gluten free biscuits using water chestnut and makhana at different levels alongwith potato powders and suggested that potato powder or makhana powder up to 50% with water chestnut powder is suitable for biscuit making particularly used by fasting people of India. Dhankar (2015) developed gluten free coconut based cookies those were comparable with the control sample for its organoleptic qualities.

Use of composite flours for biscuit making

Blends of different types of flours alongwith the wheat flour have been tried by different researchers to improve the nutritional quality of the biscuits. Sahu et al. (2015) evaluated the biscuits made from whole roasted and unroasted cereal and millet grains. The cookies were found acceptable for their physical and sensory qualities. The utilization of roasted or unroasted cereal and millet flours improved the physical properties, functional properties, sensory qualities and nutritional values of biscuits. The whole grain biscuits contain high quality soluble fiber, protein, calorific value as well as micronutrients like vitamin and minerals and can be used for feeding the targeted socio-economically poor and vulnerable groups of children and adults. El-Sharnouby et al. (2011) partially replaced wheat flour with a mixture of wheat bran and date palm fruit powder (1:1) for preparation of biscuits at different levels. The result showed that the sensory properties of the biscuits prepared using wheat flour partially replaced with a mixture of wheat bran and date palm fruit powder (30%) had acceptable organoleptic properties. Kabuo et al. (2018) processed the cocoyam

(Xanthosoma sagittifolium cv okoriko) into flour and added to wheat flour in various ratios for biscuit preparation. The results indicated that cocoyam flour can be used to replace the wheat flour upto 80% for biscuit preparation without affecting the sensory attributes adversely. Amaranth grain (*Amaranthus cruentus*) is a pseudo cereal and a good sourceof nutrients like protein, crude fiber and vitamins.

Iftikhar et al. (2015) replaced wheat flour partially with date palm paste at 10, 15 and 20% level. The organoleptic properties of the cookies prepared from the formulation using 15% date paste was highly acceptable. Mushroom is a good source of protein. Prodhan et al., (2015) explored the possibilities of using mushroom powder (MF) in biscuits preparation without impairing their acceptability. Nutrient composition and sensory investigation of biscuits incorporated with various levels of oyster mushroom (*Pleurotussajor-caju*, PSC) powder were studied. The results suggested that incorporation of mushroom powder up to 10% to replace wheat flour increased concentration of protein, dietary fiber, ash and lowering carbohydrate in biscuits with adversely affecting the organoleptic properties.

Gurung et al. (2016) utilized a blend of wheat flour and pumpkin puree for biscuit preparation. No significant decrease in diameter and thickness of the biscuits was observed however increased amount of pumpkin increased the bulk density. Pumpkin addition significantly affected the nutritional quality of the biscuits. Wheat flour and pumpkin pulp mixture in a ratio 70:30 was found best for biscuit preparation.

Öksüz and Karakaş (2016) prepared five different formulations containing buckwheat (*Fagopyrum esculentum*) flour for biscuit making. Table sugar was replaced in the formulation with carob syrup. The formulation contained purified starch as well. The results showed that the biscuit prepared had higher outer appearance scores and significantly harder texture. The biscuits were good in sensory quality. Banureka and Mahendran (2011) used wheat flour partially replaced with soy flour (from 0 to 25 %) for the production of biscuit. The nutritional composition such as protein, fat as well as energy (calorie) value of soy flour supplemented biscuits increased with progressive increase in proportion of soy flour. There was no significant difference in the sensory qualities of the soy supplemented biscuits from the control samples. 10% soybean flour incorporated biscuits were found the best among all the samples.

Husaain et al. (2018) successfully formulated multigrain biscuits by blending refined wheat flour with barley and buckwheat flours. Sindhu et al (2016) developed biscuit incorporated with defatted soya flour and carrot pomace powder to enhance the nutritional value of it and also for by-product utilization. The refined wheat flour was added with defatted soya flour in different proportion to which carrot pomace was also added. The fibre content as well as carotene content of the biscuits increased with increasing the amount of carrot pomace powder in blended flours. Similarly the protein content increased linearly with the increasing defatted soya flour in biscuits.

Sambavi et al. (2015) prepared three types of cookies using a combination of foxtail millet and wheat flour. As per the findings of sensory evaluation, sample prepared using 55% foxtail millet and 45% wheat flour sample was found highly acceptable in terms of texture, aftertaste and overall acceptability. There was no statistically significant ($p>0.05$) difference in appearance and flavor of the samples prepared. Kumar et al. (2015) developed a multigrain premix mixing whole barley, sorghum, chickpea, pea and defatted soya flour (each at 20% level) for improvement in the nutritional value of biscuits. The biscuits were prepared using this multigrain premix and wheat flour contained high amount of protein and dietary fibre. Various ratios of wheat flour and multigrain premix were tried for biscuit preparation. The biscuits prepared using multigrain mix and wheat flour blend (40:60) were found best and had 16.61% protein, 2.57% soluble fibre, and 6.67% insoluble fibre which was significantly ($p \leq 0.05$) higher than control biscuit.

Peter Ikechukwu et al. (2017) formulated cookies from whole wheat flour and date palm fruit pulp as sugar substitute. The cookies made from whole wheat – date palm in a ratio of 70:30 had the highest score in sensory analysis. Silky et al (2014) developed high fibre biscuits from the blends of refined flour and wheat bran. The results of sensory evaluation concluded that the incorporation of wheat bran in blends of refined flour up to 20% for developing biscuits was found most suitable.

Use of enzymes in bakery industry

Enzymes are widely used in bakery industry now a day. Increasing restrictions for chemical additives' usage especially in bread and similar products has significantly increased the popularity of enzymes (Cauvain and Young, 2006). Enzymes are available endogenously as well as added in wheat flour exogenously for preparation of baked products as baking aid (Di Cagno et al., 2003). Addition of enzymes help in modification of rheological behavior of dough, gas retention property and softness of bread crumb whereas for product softness in cakes and for reduction of acrylamide formation (Cauvain and Young, 2006). Addition of enzymes can be done individually or in combination so that they may act in a synergistic way for the baked good production that too at a very low level (Collar et al., 2000; Martinez-Anaya and Jimenez, 1997).

The commonly used enzymes in bakery enzymes are amylolytic enzymes such as *α-Amylases and β-Amylases and* glucoamylases. The α- and β-amylases exhibit different but complementary functions in bread preparation (Martin and Hoseney, 1991). The function of α-amylases is to break down the damaged starch particles into low molecular weight dextrins whereas β-amylase converts these dextrins in maltose. Yeast and sourdough microorganisms use maltose for fermentation (Synowiecki 2007; Goesaert et al., 2005). Addition of these enzymes also significantly improves gas retention properties of the dough after fermentation as well as reduce the dough viscosity during the starch gelatinization is observed. As a result the volume and softness of the product improve (Cauvain and Young, 2006; Goesaert et al., 2009; Poutanen 1997). The hydrolyzed starch exhibits modified retrogradation behavior. Hence the amylases are effective as anti-staling agent (Gerrar et al., 1997; Leman et al., 2009). Use of proteases reduces the mixing time of dough and decreases dough consistency assuring uniform dough and strong gluten in bread to control the bread texture and to improve the flavor (Goesaert et al., 2005, Di Cagno et al., 2003). Bombara et al. (1997) has mentioned about the application of proteases waffles for the production various products such as bread, crackers, pastries, biscuits and cookies manufacturing. Enzymes work on gluten proteins reducing its elasticity that in turn reduces the shrinkage of dough after moulding and sheeting (Cauvain and Young, 2006; Kara et al., 2005).

The most recent enzyme for bakery industry is lipase. Lipases works for improvement of dough rheology, increasing dough strength and stability so that the dough machinability can be improved (Martinez et al., 1997; Qi Si, 1997; Olesen, 2000). Lipase also leads to an increased volume resulting in more uniform and softer crumb (Qi Si, 1997). Rate of staling in baked products retards after lipase addition (Johnson, 1968; Siswoyo, 1999). Colour of the dough can be improved using lipoxygenases that bleaches fat-soluble carotenoid as a result of oxidation (McDonald, 1979; Stauffer, 1990). Lipoxygenases are also suitable for improving the mixing tolerance and dough handling properties as it oxidizes polyunsaturated fatty acids while mixing the dough (Cumbee et al., 1997). Strength of weak dough can be improved by combining sulfhydryl oxidase with glucose oxidase and xylanases (Linko et al., 1997; Faccio et al., 2012). Potassium bromate earlier used as oxidizing agent for bread making is banned in many countries because of its carcinogenic characteristic. Glucose oxidase is a suitable alternative as oxidizing agent in place of potassium bromate in bread preparation (Moore and Chen, 2006).

Packaging developments in bakery industry

Bakery and confectionary products packaging has also evolved over the time. For example, traditionally waxed paper wrappers were used for bread and biscuits but with the advancements in flexible packaging, there is a shift towards the use of different types of the flexible packaging. Bakery products may be classified in two categories on the basis of their packaging requirement:

I Low moisture bakery products (e.g. biscuits and cookies)

II Intermediate or high moisture bakery products (e.g. bread and cake)

Baked products in both the categories have different packaging requirement. Low moisture bakery products are more preferably packed in flexible packaging but the use of thermoformed plastic trays is also very common. The chances of product breaking during transportation and storage/ handling because of mechanical shock is greatly reduced using endfold style portion packs or pillow packs puffed up with gas. Various flexible packaging materials are used as pouches viz. cellophane, High-density polyethylene (HDPE) and polypropylene (PP), biaxially oriented polypropylene (OPP) etc.

Duplex oriented poly propylene (OPP) or different combinations of OPP with Polyethylene of polyethylene terepthalate (PET) are widely used for high quality products. Uses of laminates such as cellophane/ polyethylene, coated foils or metalized polyester/ polyethylene are also used. Presently composite laminates having different components with different functions is being used for biscuits. These laminates are preferred due to their various desirable properties like barrier properties against moisture and gas, ability to get heat sealed, easy printing, high production rate and overall economy (Robertson 1993; Coles and et al., 2003). Thermoformed plastic trays made of polystyrene or PVC with multiple cavities is also being used for the biscuits. The protection from vapour as well as oxygen can be done in such trays having a snap on lid, overwrapped or shrink-wrapped or sealed with a lid (Robertson 1993, 2010).

Bread and cake types of intermediate or high moisture products are marketed fresh and usually stored at ambient temperature however shelf life can be extended by refrigerated or frozen storage. The packaging material should be slightly water vapor-proof to prevent rapid drying out. If the packaging material is moisture impermeable, water will condense inside, foggy appearance and mold growth may occur.

The use of high density poly ethylene (HDPE), Low density poly ethylene (LDPE), Linear Low density poly ethylene (LLDPE), heat sealable waxed papers, semi-moisture-proof cellophanes can also be seen. Breads packed in LDPE bags with the end twisted and sealed with a strip of adhesive tape, a plastic clip or wire ties is very common now a days. Perforated LDPE bags for specialty bread to keep the crust crisper by allowing the moisture to escape from the package is also in practice (Pagani et al., 2006). Alongwith the moisture migration properties, package for cake should have sufficient rigidity to protect the cakes from any physical damage. Films of Poly Vinyl Chloride (PVC), Poly Propylene (PP), cast polypropylene (CPP), thermoformed containers of polystyrene (PS), and cellulose acetate are commonly used for cakes (Kumar and Balasubrahmanyam 1984; Robertson 1993; Coles et al., 2003).

Modified atmosphere packaging of baked products is the latest trend. The food products can be kept fresh using MAP and food safety under certain conditions may be improved (Farber et al., 1990). The recommendation for bakery products is the use of 20 to 50% CO_2 for prevention of molds growth such as *Penicillium* and *Aspergillus* alongwith 80 to 50% N_2 for increased inertia, respectively (Kotsianis et al., 2002). Usually MAP of bakery products is done with laminated or flexible films as well as semi-rigid plastics containers (Crosby 1981; Ooraikul 1991). Use of active packaging incorporating oxygen scavengers/ absorbers and ethanol can also be done for bakery items. Reduction in oxygen content by the scavenger can control the growth of molds and the oxidative changes effectively (Harima 1990; Seiler 1998). For active packaging, PVdC coated nylon, polyester, LDPE, PP, EVOH, and PS are used with some active components added. Application of ethanol has shown the effect on shelf-life of bakery items due to its anti-microbial characteristic (Powers and Berkowitz 1990; Seiler

1998). The growth of molds, yeasts, lactobacilli, and other microbial contaminants has effectively retarded by use of ethanol alongwith the reduction in the rate of staling. Ethanol vapour generator can be used for adding ethanol in the atmosphere or ethanol encapsulated in a carrier material and enclosed in plastic pouches can also be effectively used (Nielsen, 2004).

Conclusion

As per the discussion held in the chapter, it can be concluded that bakery and confectionary technology has evolved over the time. There are not only the traditional ingredients and methods those are being used for a long time but new ingredients and technologies are also available to convert the baked foods into healthier option and to improve their quality and safety simultaneously. There are still a lot of potential for research and product development in this segment.

References

Akubor, P.I. and Obiegbuna, J.E. (2014). Effect of processing methods on the quality of flour and bread from African breadfruit kernel flour. *Food Science and Quality Management*, 24, 32-41.

Alvarez-Jubete, L., Auty, M., Arendt, E.K. and Gallagher, E. (2010). Baking properties and microstructure of pseudocereal flours in gluten-free bread formulations. *European Food Research and Technology*, 230, 437-445.

Andrade, F.J.E.T., de Albuquerque, P.B.S., de Seixas, J.R.P.C., Feitoza, G.S., Barros Júnior, W., Vicente, A.A. and Carneiro-da-Cunha, M. das G. (2018). Influence of Cassia grandis galactomannan on the properties of sponge cakes: a substitute for fat. *Food & Function*, 9(4), 2456–2468. DOI: 10.1039/c7fo01864a

Baeva, M.R., Panchev, I.N. and Terzieva, V.V. (2004). Comparative study of texture of normal and energy reduced sponge cakes. *Nahrung*, 44(4), 242-6.

Bajpai, V., Shukla, P. and Shukla, R.N., (2018). Development and Quality Evaluation of Oat-Fortified Cake Rusk. *Indian Journal of Nutrition*, 5(1), 179-183

Banureka, V. and Mahendran, T. (2011). Formulation of Wheat-Soybean Biscuits and their Quality Characteristics. *Tropical Agricultural Research and Extension*, 12(2), 62-66.

Bárcenas, M.E. and Rosell, C.M. (2006). Different approaches for improving the quality and extending the shelf life of the partially baked bread: low temperatures and HPMC addition. *Journal of Food Engineering*, 72, 92-99.

Bárcenas, M.E. and Rosell, C.M. (2007). Different approaches for increasing the shelf life of partially baked bread: Low temperatures and hydrocolloid addition. *Food Chemistry, 100(4), 1594–1601.*

Basman, A. and Koksel, H. (2003). Utilization of Transgluranase use to increase the level of barley and soy flour incorporation in wheat flour breads. *Journal of Food Science*, 68(8), 2453-2460.

Ben Jeddou, K., Bouaziz, F., Zouari-Ellouzi, S., Chaari, F., Ellouz-Chaabouni, S., Ellouz-Ghorbel, R. and Nouri-Ellouz, O. (2017). Improvement of texture and sensory properties of cakes by addition of potato peel powder with high level of dietary fiber and protein. *Food Chemistry*, 217, 668–677.

Bojnanská, T., Francáková, H., Lísková, M. and Tokár, M. (2012). Legumes-The alternative raw materials for bread production. *The Journal of Microbiology, Biotechnology and Food Sciences*, 1, 876-886.

Bombara, N., Anon, M.C and Pilosof, A.M.R. (1997). Functional Properties of Protease Modified Wheat Flours. *LWT Food Science and Technology,* 30(5), 441–447

Cauvain, S. and Young, L. (2006) Ingredients and their influences. In: Cauvain S, Young L. (eds.) Baked Products. Science, Technology and Practice. Oxford: Blackwell Publishing, 72-98.

Clarke, C.I., Schober, T., Dockery, P., O'Sullivan, K. and Arendt, E.K. (2004). Wheat sourdough fermentation: effects of time and acidifi- cation on fundamental rheological properties. *Cereal Chemistry*, 81, 409–417

Coles, R., McDowell, D., Kirwan, M.J., (2003). Food Packaging Technology. Iowa: Blackwell.

Collar, C., Martinez, J.C., Andreu, P. and Armero, E. (2000). Effects of enzyme associations on bread dough performance. A response surface analysis. *Food Science and Technology International,* 6(3), 217–226.

Corsetti, A. (2012). Technology of sourdough fermentation and sour- dough appplications. In S. Cappelle, L. Guylaine, M. Gänzle, & M. Gobbetti (Eds.), Handbook on sourdough biotechnology (first ed.). Springer: Germany.

Crockett, R., Ie, P. and Vodovotz, Y. (2011). Effects of soy protein isolate and egg white solids on the physicochemical properties of gluten-free bread. *Food Chemistry,* 129(1), 84–91.

Crosby, N.T. (1981). Food packaging requirements. In: Crosby NT, editor. Food packaging materials – aspects of analysis and migration of contaminants. London: Applied Science. 9–18.

Cumbee, B., Hildebrand, D.F. and Addo, K. (1997). Soybean flour lipoxygenase isozymes effects on wheat flour dough rheological and breadmaking properties. *Journal of Food Science,* 62(2), 281-283

De Angelis, M., Cassone, A., Rizzello, C.G., Gagliardi, F., Minervini, F., Calasso, M., Di Cagno, R., Francavilla, R. and Gobbetti, M. (2010). Mechanism of degradation of immunogenic gluten epitopes from Triticum turgidum L. var. durum by sourdough lactobacilli and fungal proteases. *Applied and Environmental Microbiology*, 76, 508-518.

Defloor, I., Nys, M. and Delcour, J.A. (1993). of the Breadmaking Potential of Wheat Flour. *Cereal Chem, 70*(5), 526-530

Demirkesen, I., Mert, B., Sumnu, G. and Sahin, S. (2010). Utilization of chestnut flour in gluten-free bread formulations. *Journal of Food Engineering,* 101, 329-36.

Dhankhar, P. (2013). A Study on Development of Coconut Based Gluten Free Cookies. *International Journal of Engineering Science Invention,* 2 (12), 10-19

Dhingra, S. and Jood, S. (2002). Physico-chemical and nutritional properties of cereal-pulse blends for bread making. *Nutrition and health, 16*(3), 183-194

Dhingra, S. and Jood, S. (2001) Organoleptic and Nutritional Evaluation of Wheat Breads Supplemented with Soyabean and Barley Flour. *Food Chemistry*, 77, 479-288.

Di Cagno, R., De Angelis, M., Auricchio, S., Greco, L., Clarke, C., De Vincenzi, M., Giovannini C., D'Archivio, M., Landolfo, F., Parrilli, G., Minervini, F., Arendt, E. and Gobbetti, M. (2004). Sourdough bread made from wheat and nontoxic flours and started with selected lactobacilli is tolerated in celiac sprue patients. *Applied and Environmental Microbiology.* 70, 1088 -1096.

Di Cagno, R., De Angelis, M., Corsetti, A., Lavermicocca, P., Arnault, P., Tossut, P., Gallo, G. and Gobbetti, M. (2003). Interactions between sourdough lactic acid bacteria and exogenous enzymes: effects on the microbial kinetics of acidification and dough textural properties. *Food Microbiology,* 20(1), 67–75. DOI: 10.1016/s0740-0020(02)00102-8

Diez-Sánchez, E., Llorca, E., Quiles, A. and Hernando, I. (2018). Using different fibers to replace fat in sponge cakes: In vitro starch digestion and physico-structural studies. *Food Science and Technology International,* 24(6), 533–543.

Doulia, D., Katsinis, G. and Mougin, B. (2000). Prolongation of the microbial shelf life of wrapped part baked baguettes. *International Journal of Food Properties,* 3(3), 447–457.

Doxastakis, G., Zafiriadis, I., Irakli, M., Marlani, H. and Tananaki, C. (2002). Lupin, soya and triticale addition to wheat flour doughs and their effect on rheological properties. *Food Chemistry*, 77(2), 219-227

El-Sharnouby, G.A., Aleid, S.M. and Al-Otaibi, M.M. (2012). Nutritional quality of biscuit supplemented with wheat bran and date palm fruits (Phoenix dactylifera L.). *Food and Nutrition Sciences*, *3*(03), 322.

Faccio, G., Flander, L., Buchert, J., Saloheimo, M. and Nordlund, E. (2012). Sulfhydryl oxidase enhances the effects of ascorbic acid in wheat dough. *Journal of Cereal Science,*55(1), 37-43

Farber, J.M., Warburton, D.W., Gour, L., Milling, M. (1990). Microbiological quality of foods packaged under modified atmospheres. *Food Microbiology*, 7:327–34.

Fernandes, S.S. and Salas-Mellado, M.L. (2017). Addition of chia seed mucilage for reduction of fat content in bread and cakes. *Food Chemistry,* 227, 37-44

Gänzle, M.G. (2014). Enzymatic and bacterial conversions during sour- dough fermentation. *Food Microbiology*, 37, 2–10.

Gerrard, J.A., Every, D., Sutton, K.H. and Gilpin, M.J. (1997). The role of maltodextrins in the staling of bread. *Journal of Cereal Science*, 26(2), 201–209

Giami, G.Y., Amasisi, T. and Ekiyor, G. (2004). Comparison of bread making properties of composite flour from kernels of roasted and boiled African bread fruit (Treculia africana) seed. *Journal of Material Research*, 1(1): 16-25

Goesaert, H., Brijs, K., Veraverbeke, W.S., Courtin, C.M., Gebruers, K. and Delcour, J.A. (2005). Wheat flour constituents: how they impact bread quality, and how to impact their functionality. *Trends in Food Science and Technology,* 16(1-3), 12–30

Goesaert, H., Slade, L., Levine, H. and Delcour, J.A. (2009). Amylases and bread firming – an integrated view. *Journal of Cereal Science,* 50(3), 345–352

Gómez, M., Oliete, B., Rosell, C.M., Pando, V. and Fernández, E. (2008). Studies on cake quality made of wheat–chickpea flour blends. *LWT-Food Science and Technology*, *41*(9), 1701-1709.

Gurung, B., Ojha, P. and Subba, D. (2016). Effect of mixing pumpkin puree with wheat flour on physical, nutritional and sensory characteristics of biscuit. *Journal of Food Science and Technology Nepal*, *9*, 85-89

Hammes, W.P., Ganzle, M.G. (1998). Sourdough breads and related products. In: Woods BJB, ed. Microbiology of Fermented Foods Vol. 1. London: Blackie Academic/Professional, 199-216

Harima, Y. (1990). Free oxygen scavengers. In: Kadoya T, editor. Food packaging. New York: Academic Press. 477–91.

Hussain, A., Kaul, R. and Bhat, A. (2018). Development and Evaluation of Functional Biscuits from Under utilised Crops of Ladakh. *International Journal Current Microbiological Application and. Science*, 7(3), 2241-2251

Iftikhar, F., Kumar, A., Altaf, U. (2015). Development and Quality Evaluation of Cookies Fortified with Date Paste (*Phoenix dactylifera* L), *International Journal of Science, Engineering and Technology*, 3 (4), 975-978.

Johnson, R.H. and Welch, E.A. (1968). Baked goods dough and method. US Patent Application; 1968. US 3,368,903.

Kabuo, N.O., Alagbaoso, O.S., Omeire, G.C., Peter-Ikechukwu, A.I., Akajiaku, L.O. and Obasi, A.C. (2018). Production and Evaluation of Biscuits from Cocoyam (*Xanthosoma Sagittifolium* Cv Okoriko)-Wheat Composite Flour. *Research Journal of Food and Nutrition*, *2*(2), 53-61.

Kara, M., Sivri, D. and Koksel, H. (2005). Effects of high protease-activity flours and commercial proteases on cookie quality. *Food Research International,* 38(5), 479–486.

Kaur, H. and Sharma, S. (2018). Development and nutritional evaluation of cake supplemented with pumpkin seed flour, *Asian Journal of Dairy & Food Research*, 37(3), 232-236.

Khalil, A.H., Mansour, E.H. and Dawoud, F.M. (2000). Influence of malt on rheological and baking properties of wheat–cassava composite flours. *LWT-Food Science and Technology*, *33*(3), 159-164.

Kotsianis, I.S., Giannou, V., Tzia, C. (2002). Production and packaging of bakery products using MAP technology. *Trends in Food Science and Technology,* 13, 319–24.

Krupa-Kozak, U., Baczek, N. and Rosell, M.C. (2013). Application of dairy proteins as technological and nutritional improvers of calcium-supplemented gluten-free bread. *Nutrient*, 5, 4503-4520.

Kumar, K. A., Sharma, G. K., Khan, M. A., Semwal, A.D. (2015). Optimization of Multigrain Premix for High Protein and Dietary Fibre Biscuits Using Response Surface Methodology (RSM). *Food and Nutrition Sciences*, 6, 747-756.

Kumar, K.R. and Balasubrahmanyam, N. (1984). Plastics in packaging. S.A.P. Vaidya (edi.), Indian Institute of Packaging. 319–341.

Leman, P., Goesaert, H. and Delcour, J.A. (2009). Residual amylopectin structures of amylase-treated wheat starch slurries reflect amylase mode of action. *Food Hydrocolloids,* 23(1), 153–164

Leuschner, R.G.K, O'Callaghan, M.J.A and Arendt, E.K. (1999). Moisture distribution and microbial quality of part baked breads as related to storage and rebaking conditions. *Journal of Food Science,* 64, 543-546.

Linko, Y-Y., Javanainen, P. and Linko, S. (1997). Biotechnology of bread baking. *Trends in Food Science and Technology*, 8(10), 339-344.

López, A.C.B., Pereira, A.J.G., Junqueira, R.G. (2004) Flour mixture of rice flour, corn and cassava starch in the production of gluten-free white bread. *Brazilian Archives of Biology and Technology*. 47:63–70.

Luz Fernandez, M.A.R.I.A. and Berry, J.W. (1989). Rheological properties of flour and sensory characteristics of bread made from germinated chickpea. *International Journal of Food Science & Technology*, *24*(1), 103-110.

Makhoul, S., Romano, A., Capozzi, V., Spano, G., Aprea, E., Cappellin, L., Benozzi, E., Scampicchio, M., Märk, T.D., Gasperi, F., El-Nakat, H., Guzzo, J. and Biasioli, F. (2015). Volatile compound production during the bread- making process: effect of flour, yeast and their interaction. *Food and Bioprocess Technology*, 8, 1925–1937

Man, S.M., Paucean, A. and Muste, S. (2014), Preparation and Quality Evaluation of Gluten-Free Biscuits. *Bulletin UASVM Food Science and Technology,* 71(1), 38-44

Manini, F., Casiraghi, M.C., Poutanen, K., Brasca, M., Erba, D. and Plumed-Ferrer, C. (2016). Characterization of lactic acid bacteria isolated from wheat bran sourdough. *LWT - Food Science and Technology*, 66, 275-283.

Manley, D. (2000). Manley's technology of biscuits, crackers and cookies. Woodhead Publishing

Marco, C. and Rosell, M.C. (2008). Bread making performance of protein enriched, gluten-free breads. *European Food Research Technology,* 227, 1205-13.

Martin, M.L. and Hoseney, R.C. (1991). A mechanism of bread firming. II role of starch hydrolyzing enzymes. *Cereal Chemistry,* 68(5), 503–509

Martinez-Anaya, M.A. and Jimenez, T. (1997). Rheological properties of enzyme supplemented dough's. *Journal of Texture Studies,*28(5), 569–583

McDonald, C.E. (1979). Lipoxygenase and lutein bleaching activity of durum wheat semolina. *Cereal Chemistry*, 56(2) 84–89.

Menon, L., Majumdar, S.D. and Ravi, U. (2014). Mango *(Mangifera indica* L.) kernel flour as a potential ingredient in the development of composite flour bread. *Indian Journal of Natural Products and Resources*, 5(1),75-82.

Miller, R.A., Dann, O.E., Oakley, A R., Angermayer, M.E. and Brackebusch, K.H. (2017). Sucrose replacement in high ratio white layer cakes. *Journal of the Science of Food and Agriculture*, 97(10), 3228–3232.

Minarro, B., Albanell, E., Aguilar, N., Guamis, B. and Capellas, M. (2014). Effect of legume flours on baking characteristics of gluten-free bread. *Journal of Cereal Science,* 56: 476-8.

Mishra, A., Devi, M. and Jha, P. (2014). Development of gluten free biscuits utilizing fruits and starchy vegetable powders. *Journal of Food Science and Technology,* 52(7), 4423–4431.

Mohammed, I., Ahmed, A.R. and Senge, B. (2012). Dough rheology and bread quality of wheat–chickpea flour blends. *Industrial Crops and Products*, *36*(1), 196-202.

Moore, M.M. and Chen, T. (2006). Mutagenicity of bromate: Implications for cancer risk assessment. *Toxicology,* 221(2-3), 190–196.

Moraes, É.A., Dantas, M.D.S., Morais, D.D.C., Silva, C.O.D., Castro, F.A.F.D., Martino, H.S.D. and Ribeiro, S.M.R. (2010). Sensory evaluation and nutritional value of cakes prepared with whole flaxseed flour. *Food Science and Technology*, *30*(4), 974-979.

Nielsen, P.V. (2004). Packaging, quality control, and sanitation of bakery products. In: Hui, Y.H., Meunier- Goddik, L., Hansen, A.S., Josephson, J., Nip, W-K., Stanfield, P.S., Toldra F, editors. Handbook of food and beverage fermentation technology. New York: Marcel Dekker. Chapter 43.

Noor Aziah, A. A., Lee Min, W. and Bhat, R. (2011). Nutritional and sensory quality evaluation of sponge cake prepared by incorporation of high dietary fiber containing mango (Mangifera indicavar. Chokanan) pulp and peel flours. *International Journal of Food Sciences and Nutrition,* 62(6), 559–567.

Öksüz, T. and Karakaş, B. (2016). Sensory and textural evaluation of gluten-free biscuits containing buckwheat flour. *Cogent Food & Agriculture*, *2*(1), 1178693

Olesen, T., Qi Si, J. and Donelyan, V. (2000). Use of lipase in baking. US Patent Application. US 6110508.

Ooraikul, B. (1991). Modified atmosphere packaging of bakery products. In: Ooraikul B, Stiles ME, editors. Modified atmosphere packaging of food. Chichester: Ellis Horwood. p 38–114.

Pagani, M.A., Lucisano, M., Mariotti, M. and Limbo, S. (2006). Influence of packaging material on bread characteristics during ageing. *Packaging Technology Science*, 19:295–302.

Păucean, A. and Man, S. (2013). Influence of defatted maize germ flour addition in wheat: maize bread formulations. *Journal of Agroaliment Process Technology*, *19*(3), 298-304.

Peressini, D., Pin, M. and Sensidoni A. (2011). Rheology and breadmaking performance of rice-buckwheat batters supplemented with hydrocolloids. *Food Hydrocolloid*, 25, 340-9.

Peter Ikechukwu, A., Okafor, D. C., Kabuo, N. O., Ibeabuchi, J.C., Odimegwu, E. N., Alagbaoso, S. O. Njideka, N.E. and Mbah, R. N. (2017). Production and Evaluation Of Cookies From Whole Wheat And Date Palm Fruit Pulp As Sugar Substitute. *International Journal of Advancement In Engineering Technology, Management and Applied Science*, 4(4), 1-31.

Pollard, N.J., Stoddard, F.L., Popineau, Y., Wrigley, C.W. and MacRitchie, F. (2002). Lupin flours as additives: dough mixing, breadmaking, emulsifying, and foaming. *Cereal Chemistry*, *79*(5), 662-669

Poutanen, K. (1997). Enzymes: An important tool in the improvement of the quality of cereal foods. *Trends in Food Science and Technology,*8(9), 300-306.

Powers, E.D., Berkowitz, D. (1990). Efficacy of an oxygen scavenger to modify the atmosphere and prevent mold growth on meal, ready-to-eat pouched bread. *Journal of Food Protection*, 53:767–70.

Prodhan, U.K., Linkon, K.M.M.R., Al-Amin, M.F. and Alam, M.J. (2015). Development and quality evaluation of mushroom (pleurotussajor-caju) enriched biscuits. *Emirates Journal of Food and Agriculture*, 27(7), 542–547.

Psimouli, V. and Oreopoulou, V. (2012). The effect of alternative sweeteners on batter rheology and cake properties. *Journal of the Science and Food Agriculture*, 92(1), 99-105.

Psimouli, V. and Oreopoulou, V. (2013). The effect of fat replacers on batter and cake properties. *Journal of Food Science,* 78(10), C1495-C1502.

Puranik, D.B. and Gupta, S.K. (2017). Development of egg-less cake using whey protein concentrate as egg substitute. *International Journal of Science, Environment and Technology*, *6*(4), 2343-2352

Qi Si, J. (1997). Synergistic effect of enzymes for bread making. *Cereal Foods World,* 42(10), 802-807

Ribotta, P.D., Arnulphi, S.A., León, A.E. and Añón, M.C. (2005). Effect of soybean addition on the rheological properties and bread making quality of wheat flour. *Journal of the Science of Food and Agriculture*, *85*(11), 1889-1896.

Robertson, G. L. (2010). Food packaging and shelf life: A practical guide. Florida: CRC Press. Chapter 1.

Robertson, G.L. (1993). Food packaging: Principles and practice of packaging and converting technology. New York: Marcel Dekker. Chapters 1, 2, 18.

Rodríguez-García, J., Puig, A., Salvador, A. and Hernando, I. (2012). Optimization of a sponge cake formulation with inulin as fat replacer: structure, physicochemical, and sensory properties. *Journal of Food Science,* 77(2), C189-97

Rosell, C.M. and Gómez, M. (2007). Frozen Dough and Partially Baked Bread: An Update. *Food Reviews International,* 23(3), 303–319.

Sadowska, J., Błaszczak, W., Fornal, J., Vidal-Valverde, C. and Frias, J. (2003). Changes of wheat dough and bread quality and structure as a result of germinated pea flour addition. *European Food Research and Technology*, *216*(1), 46-50

Sahu, U., Prasad, K., Sahoo, P., Sahu, B.B., Sarkar, P.C. and Prasad, N. (2015). Biscuit making potentials of raw and roasted whole grain flours: Cereals and millets. *Asian Journal of Dairy and Food Research*, *34*(3), 235-238.

Sakr, A.M. and Hussien, H.A. (2017). A Nutritional quality of gluten free biscuits supplemented with sweet chickpeas and date palm powder. *International Journal of Food Science and Nutrition*, 2(1), 128-134

Salovaara, H. (1988). Lactic acid bacteria in cereal-based products. In: Salminen S, von Wright A, ed. Lactic acid bacteria— Microbiology and functional aspects. New York, NY: Marcel Dekker, 115-137.

Sambavi, A., Sabaragamuwa, R.S. and Suthakaran, R. (2015) Development of Cookies Using a Combination of Foxtail Millet and Wheat Flour. *International Journal of Scientific & Technology Research*, 4(10), 294-295.

See, E.F., Noor Aziah, A.A. and Wan Nadiah, W.A. (2007). Physico-chemical and sensory evaluation of breads supplemented with pumpkin flour. *ASEAN Food Journal*, *14*(2), 123

Seiler, D.A.L. (1998). Bakery products. In: Blakistone BA, editor. Principles and applications of modified atmosphere packaging of foods. 2nd ed. London: Blackie Academic & Professional. 135–57.

Seth, K. and Kochhar, A. (2018), Nutritional Assessment of Healthy Cakes Developed Using Partially Defatted Peanut Flour, *Chemical Science Review and Letters*, 7(25), 244-249

Shahin, F.M. and Sakr, A.M. (2016). Technological and Nutritional Evaluation of Biscuits Fortified Amaranths. *Middle East Journal of Applied Science*, 6(3), 449–459

Sharma, S., Bajwa, U. and Nagi, H.P.S. (1999). Rheological and baking properties of cowpea and wheat flour blends. *Journal of the Science of Food and Agriculture*, *79*(5), 657-662

Shin, D-J., Kim, W. and Kim, Y. (2013). Physicochemical and sensory properties of soy bread made with germinated, steamed, and roasted soy flour. *Food Chemistry,* 141(1), 517–523.

Sindhu, L.H., Sh, S., Harshavardhan, K., Mounika, B., Kalyani, D. and Pavankumar, N.S. (2016). Development of biscuit incorporated with defatted soya flour and carrot pomace powder. *IOSR Journal of Environmental Science, Toxicology and Food Technology,* 10(3), 27-40.

Singh, T. and Chaudhary, N. (2016). Development of high fibre cake using gram flour. *International Journal of Information Research and Review,* 3, 2513-2515.

Siswoyo, T.A., Tanaka, N. and Morita, N. (1999). Effects of lipase combined with α-amylase on retrogradation of bread. *Food Science and Technology Research,* 5(4), 356-361

Stauffer, C.E. (1990). Enzymes. In: Stauffer CE. (ed.) Functional Additives for Bakery Foods. New York: Van Nostrand Reinhold, 148-152.

Stauffer, C.E. (1993) Frozen dough production. In: Kamel B.S., Stauffer C.E. (eds) Advances in Baking Technology. Springer, Boston, MA.

Synowiecki, J. (2007). The Use of Starch Processing Enzymes in the Food Industry. In: Polaina J, MacCabe AP. (eds.) Industrial Enzymes. Sctructure, Function and Applications. Dordrecht: Springer, 19-34.

Tharshini, G., Sangwan, V. and Suman, (2018). Organoleptic and chemical characteristics of soybean and pomegranate peel powder supplemented cakes. *Journal of Pharmacognosy and Phytochemistry,* 7(2), 35-39.

Veluppillai, S., Nithyanantharajah, K., Vasantharuba, S., Balakumar, S. and Arasaratnam, V. (2010). Optimization of bread preparation from wheat flour and malted rice flour. *Rice science*, 17(1), 51-59.

Wronkowska, M., Haros, M. and Soral-Smietana, M. (2013). Effect of starch substitution by buckwheat flour on gluten-free bread quality. *Food Bioprocess Technology*, 6, 1820-1827.

Ziobro, R., Witczak, T., Juszczak, L. and Korus, J. (2013). Supplementation of gluten-free bread with non-gluten proteins. Effect on dough rheological properties and bread characteristic. *Food Hydrocolloids,* 32(2), 213–220.

3

Recent Advancements in Sugar, Honey, Salt and Jaggery Processing

Pranya Prashant[1], *Prasad Rasane*[1*], *Arvind Kumar*[2], *Jyoti Singh*[1]
Sawinder Kaur[1]

[1]*Department of Food Technology and Nutrition, Lovely Professional University, Phagwara, Punjab*
[2]*Department of Dairy Science and Food Technology, Banaras Hindu University, Varanasi, Uttar Pradesh*

Advancements on sugar processing

India has popped out as one of the largest sugar processing country throughout the world. Through the collaboration of research and development carried out in the field of sugar processing, this industry has witnessed the vast changes in term of technology. Important considerations such as productivity improvement and cost reduction were the main objectives in the field of sugar processing. Sulphur dioxide was earlier used but afterwards lime treatment in clarified juice helped in the processing of white sugar from sugar cane. Since then the continuous progress and improvement has been made and advanced techniques are introduced for manufacturing of sugar with modern tools.

Ramjeawon (2004) reported that farmers and agriculture workers grow sugar cane and beet root for sugar production,providing about 20 kg per capita and that is equivalent to almost 13 percent energy requirement. Fechter et al.(2001)reported that production of the white sugar was focused more on juices rather than molasses which were taken as the raw material. Various processes such as ultra-filtration, de-mineralization and de-colorization were used for the production of white sugar. The traditional refining process wascheaper, and effective in terms of purification and juice purification was the method used for the processing of white sugar directly from the sugar cane. Crystallization was one of the process which is used for increasing the recovery of sucrose and it reduces the molasses production. Chromatography was used in the sugar industry for the separation of sucrose. Another method was ion exchange demineralization which was used in the sugar industry for the removal of minerals from the liquid sugar. Later adsorption has been incorporated to be used de-odorization and de-pigmentation. Membrane filtration is latest technique used in the process which was used for maintaining the quality of white sugar by removing unwanted particle from the juice. Velásquez et al.,(2018) reported that thermal processes are also used for the production of sugar from sugar cane. Non centrifugal cane sugar processing is used for this purpose and the product is obtained by evaporating the sugarcane which has the unique flavor and aroma. Later the product is used as the sweetener and has a traditional characteristics flavor and optimum nutrition.

Sugar based products

The recent updates in sugar based products are presented in Table 1. Zumbe et al., (2001) reported that the sugar products in terms of confectioneries are popularly consumed around the world, with sophisticated modifications and different attributes including sweetness, aroma, flavor, texture and mouth-feel. Confectionery is the term given to a product that uses sugar in one form or another as a major ingredient. These forms include dextrose, sucrose, glucose, lactose. Commonly sucrose and glucose are chosen because of their easy availability and cheaper expense to the consumer. Souhail et al.(2009) reported that dates are incorporated in the production of jams and have very good sensory quality and also has high commercial value. Dates have hard texture and it consists of sugar and fiber that have high value components and they have many values added applications. Jam is used as the preservative for fruits. Dates are incorporated with various ingredients for the preparation of jams which helps in the value addition of the date fruit.

Table 1: Recent updates in sugar based products

Product	Nutraceutical Inventory	Health Benefits	References
Jackfruit candy	Incorporation of jack-fruit.	It helps in curing cancer. It helps in proper digestion.	Shrikant et al,2012
Sacred figs	Incorporation of figs which is commonly known as peepal.	It helps in the treatment of asthama.	Isha et al, 2015
Olive candy	Incorporation of olive.	It helps in weight loss. It reduces the risk of breast cancer.	Abdellaui et al, 2018
Aonla candy	Incorporation of aonla.	It acts as a good appetizer. It helps in digestion. It maintains the metabolism rate.	Murlidhar et al, 2016
Lollipop	Incorporation of licorice extract.	It acts as antimicrobial agent. It reduces the cavity causing bacteria in body.	Chu-hong et al, 2001
Date jam	Incorporation of date fruit.	It acts a dietary fiber which is a natural occurring source of prebiotics.	Souhail et al, 2008
Marmalades	Incorporation of rose hip fruit.	It acts as an antioxidant.	Oktay et al, 2012
Barfi	Incorporation of star fruit.	It helps in preventing cancer.	Jorge et al, 2010

Eonard et al.(1998) reported that the candy with the essence of some medication property is very useful for children and it provides very safe and convenient means of administrating medicine. Prosenjit et al. (2014) reported that the fruits are basically preserved using the process of candy making. These preserves include jelly, marmalades fruits bar is prepared as the sugar-based products.Sonneratia apetala fruit was used for the preparation of vitamin C rich jelly.Other common types of sugar-based fruit product consumed are fruit juice which is liquid parts of fruit, obtained by squeezing or crushing the fruit. The juice can be clear or turbid depending upon the need of consumption. It can be concentrated or later water is added for maintaining the quality factor of the fruit juice depending upon the suitability. Sugar based products such a jelly is the semisolid part of the fruit product made up of the fruit juice. It should have certain property like it should be crystal clear and should be free from any residual particle and it should be stiff enough to hold its proper shape. Other type of product is marmalades in which the peel of fruits is added along with the juices. Basically only citrus fruit are used for the preparation of marmalades and the peel added are boiled first before adding to remove its bitterness. Another sugar-based product is the fruit bar, which has the significant properties of cookies. It is fortificated with the chopped fruits along with some other ingredients which includes sugar, milk powder hydrogenated fats and citric acid. Pectin plays an important role in preparation of sugar based products , it is used as the thickening agent or gelling agent in certain products such as jam, jelly and other similar products.

Nutraceutical inventory in sugar based products

Nutraceutical inventory in sugarbased products enhances its functional property. Incorporation of functional raw materials makes it effective in all terms. Isha et al., (2015). Sacred figs which is commonly known as peepal which belongs to the family of *moraceae* are used for the preparation of hard candy. The nutritional profile of sacred figs is excellent as it is rich source of dietary fiber, proteins, calcium, magnesium, phosphorus and it also contains very less amount of fat. Along with these nutritional properties it also has medicinal effect which makes it moreeffective in terms of consumption. Incorporation of sacred figs for the preparation of candy is more effective as it has such nutraceutical properties which makes it very worthy, like it contains, beta caryophyllene, alpha terpinine, limonine and such effective components.

Abdellaoui et al., (2018) reported that the other such sugar-based products include the normal candy prepared from date, olive and carob fruit. The olive and date were used in the form of pastes. These fruits wereused as they high nutritional value and also have medicinal property which makes them appropriate for nutraceutical invention in final product. Murlidhar et al.,(2016) reported that the incorporation of aonla for the preparation of candy reduces its totalcalorie count and increases the overall acceptability. They reported that best anola candy was made by the aspartame sugar and found to be good for diabetic people. It is also regarded as the richest source of vitamin C which is good for teeth, gums, eye sight and digestion.

Advancements on honey processing

Subramania et al., (2007) reported that honey is the natural biological products which is produced from the nectar and is highly beneficial for the human being for the food as well as medicinal purpose.

The processing techniques of honey includes thermal processing, microwave heating, infrared heating, ultrasound processing and membrane processing. Thermal processing is commonly used for the elimination of microorganism. Rapid method for the yeast reduction is microwave heating along with least thermal damage. For shorter duration of time, infrared processingis used. Another method used is membrane processing which is athermal process, used for the complete removal of yeast cell from honey. For the production of enzyme-enriched honey the techniques used are microfiltration and ultrafiltration.

Thermal processing of honey

The main drawbacks in honey processing is the quality deduction due to the fermentation. It has been reported that the honey contains more than 20% pf moisture, because of that it influences the rate of fermentation. granulation and change in flavor of honey. The chances of fermentation are high in case of unprocessed honey. For the prevention of fermentation in honey, the honey is first processed in form of heat and then is stored. The processing with heat, reduces the risk of microorganism which are responsible for spoilage and reduces the moisture level that retards the fermentation process.

Conventional processing of honey

Processing of honey with the conventional techniques involves the preheating at 40˚C followed by straining, clarification/filtration, and it also involves the indirect heating of filtered honey at 60-65˚C for 25-30 minutes at tubular heat exchanger. This step is followed by the rapid cooling of honey which is done for the protection of its natural colors, flavors, enzyme contents and other biological substance. It has been reported that the storage condition of honey plays an very important role in quality parameters. It was seen that the color effect was clearly visible by storing the honey at temperature 40˚C.

Microwave heat processing of honey

The microwave heating in food industry has numbers of application which includes tempering, blanching, drying and pasteurization. Heating in microwave will be more effective by increasing the moisture content. Water is the major absorber of heat in microwave processing, so it greatly influences the presence of water in food. Honey contains small amount of water and large amount of dissolved sugar so microwave way of processing is very effective. Ghazali et al., (1994) reported that the storage study of starfruit honey was conducted and it showed that the temperature which requires very short time to reach the desired processing condition had very little effect on chemical property. Later honey as heated at 71° C using microwave oven and kept for storage at two different temperature, one at room temperature and other at 4°C for 16 weeks. Before storage and during the storage time, the physicochemical properties of both heated and unheated were measured and it was found that the spoilage were seen in the unheated honey and heated one was not spoiled, just irrespective of the storage condition. Color became very dark of both unheated and heated honey, however the sample that stored 4°C were lighter in color as compared to that stored in room temperature.

Infrared heat processing of honey

It is becoming very popular because of its significant savings of energy as compared to other processing techniques such as thermal processing. It consists of the hybrid system with conductive and convective heating source. High rates of infrared heaters are used which provides the input in forms of energy to the material surface. The radiant heat flux penetrates the material into the depth and this phenomenon depends upon the nature of material and wavelength of incident material. This processing is very much effective in honey formation.

Membrane processing of honey

It is the athermal process and can be used for an alternative for the thermal processing. It is difficult to destroy all the microorganism by using thermal processing and so it that case this method of processing is applied. This type of processing techniques is very much effective as it does not give any cloudiness and granular types texture in the final product. It has very good effect on the quality characteristics of the product.

Honey based products

Table 2 depicts the recent updates in honey based products. Katke et al., (2018) reported that there are many honey based products are available, one such products in candy with the incorporation of aonla fruit. In this candy sugar were replaced by the honey and then it was used for the preparation of candy. Candy can be considered as the very good value-added product and can be consumed by targeting certain age group. Umesh et al., (2008) reported that other type of product prepared from honey is honey powder which includes different types of processing techniques for the preparation of powder. Process used for the preparation of honey powder includes spray drying and drum/roller drying process. For the preparation of honey powder through spray drying technique, the waxy starch is added to honey diluted with water. The temperature is maintained at 140-150° C. The fiber content found in the honey powder was above 23%. Dextrin, maltose and anti caking agent are mixed with honey during spray drying technique and temperature is maintained at 115-125°C.Pulverizing technique is also used for the production of honey powder. Dextrin, lactose and starch ar added in the honey for the production of honey powder. Cyclodextrin is the functionl additive added for the production of honey powder.The major constraint in the preparation of the honey-based powder is stickiness of the final product. Syed et al.,(2015) reported that the spreads from honey are prepared and it is very good source of dietary fiber. Chandegara et al.,(2018) reported that the different types of chocolate and energy bars are prepared from honey. Ingrediens used for the preparartion of chocolate bars were dates,oats, rasins, rice crispies, nuts, honey, chocolate, sunflower seeds. After the preparation of the energy bar it was found that the these ingredients were present in certain proportion that were,oats(30%),honey (15%), dark chocolate (14%), nuts (10%), sunflower seeds (10%), dates (8%), rice crispes (8%) and rasins (5%).

Table 2: Recent updates in honey based products

Product	Nutraceutical Inventory	Health Benifits	References
Spreads	Incorporation of carob flour.	It is a good source of dietary fiber.	Syed et al,2015 Subramanian,2007
Candy	Incorporation of anola.	It has high energy value.	Katke et al, 2018
Dark chocolate	Incorporation of milk.	It lowers the blood glucose level.	Nurul Zaizuliana Rois Anwar et al,2018
Bar	Incorporation of nuts(Almond, Cashew)	It prevents hypohydration.	Chandegara et al,2018
Flakes	Incorporation of sucrose syrup.	It has the therapeutic property which leads for preventing dementia.	Hebbar et al,2008
Bread	Incorporation of honey in the form of powder helps in improving the dough quality including its sensory characteristics.	It act as an anti cancer agent. It helps in preventing nausea.	Sampath et al,2010. Qunyi et al,2010

These energy bars and chocolate are used for providing instant energy. These products are useful for certain target groups who requires energy in large amount or who does exercise and workouts.Various other types of honey-based products are bars, flakes, chocolates, breads which are available in the market.

Nutraceutical inventory in honey

Syed et al., (2015) reported that the various nutraceutical inventory has been developed in honey which makes the honey product more nutraceutical in nature. Products like honey spreads with the incorporation of carob flour are used which makes it rich in nutraceutical aspects and hence enhance its functional property. Carob flour is very good source of dietary fiber, which helps in reducing the problems like constipation in our body. Katke et al.,(2018) reported that the candieswere prepared from the honey with the incorporation of anola fruit which provides energy in our body and helps in maintaining energy level in our body. Nurul et al, (2018) reported that the dark chocolate were prepared from honey with the incorporation of milk powder which has significant effect in maintaining the blood glucose level in our body. Chandegara et al., (2018) reported that the bars prepared from honey with the incorporation of nuts including almonds and cashews which helps in preventing dehydration in our body. Umesh et al., (2008) reported that the product like flakes are also prepared by honey with the fortification of sucrose syruph which has many therapeutic property which helps in preventing dementia. Sampath et al., (2010) reported that the honey breads are also prepared with the incorporation of powder honey which further acts as an anti-cancer agent for the people.

Anti-microbial activity of honey

Liyanage et al., (2017) reported that honey has antimicrobial properties from various bacterial species including, bacillus antrcis, corneybacterium diptheria, haemophillus influenzae. Honey has various antiviral effect also which helps it in showing various functional properties against various types of diseases and lesions. It has been seen that the application of honey was very much effective in the management antimicrobial diseases including skin infections.

Anti-fungal effect of honey

Liyanage et al., (2017) reported that the honey has some anti-fungal effect also against aspergillus, penicillium. Various types of herbal lotions and creams are prepared from honey which further helps in preventing various types of diseases including lesions skin infection, and various other types of skin problems.

Updates on salt processing

Edel et al., (2008) reported that the salt is one of the most ancient ingredients used in the processing and manufacturing of food process. Salt is basically added during the cooking or can be added from above. Salt is basically the combination of sodium and chloride. Among this up to 40% of molecules are consist of sodium. The very common and ancient method of salt processing is production of salt from the sea water. For the production of salt, the sea water is being evaporated. This evaporation process involves three steps, single system, double system and multi-pond system.

Single system

In this type of processing system, the salt of very low grade is produced. This process is the batch process which further reduced the production cost.

Double system

The second system is known as the double system in which the evaporation basin is divided into two parts, the first basin is known as the nurse pond which is used for the production of sodium and the other part is known as crystallizer.

Multi-pond system

In this system the nurse pond is divided into various types of interconnected basins. In this type of system the three parts are utilized step by step for the production of saltbased.

Salty products updates

The recent updates on salt based products are depicted in Table 3. Ibrahim et al., (2009) reported that various types of salt-based products are available in the market. The fish protein concentrates were added upto 5%. It has very good water holding capacity and it increases the protein content in the product especially the essential amono acids. In this for the fish protein concentrates, the water and oil is removed from the fish and hence it increses the concentration of protein and other nutrients. For the preparation of product, five different protein concentrates were used from diffenrt sources such as sunflower seeds, soyabean,lupine, rice bran and fish. These ingredients were added with the white bread at 5 and 10% level. Zoinon et al., (2017) reported that the products like soy chilly sausages are produced which are processed by addition of natural preservatives. Sodium benzoate was added as the preservative for the preservation of product. It was found that addition of sodium benzoate can extend the shelf life of the product upto 1 year without effecting the taste and quality of the product. Rodge et al., (2018) reported that the instant upma mix were prepared by the addition of foxtail millet and garden cress seeds. Different proportions were tried and it was found that 75:10 were acceptable.It was found that the overall composition of final upma mix were, moisture content (6.15-7.62), protein content (11.30-13.84%), fat content (7.30-16.80%), fiber content (3.90-4.31%), ash content (2.89-4.43%), carbohydrate (55.16-64.07%).

Table 3: Recent updates in salt based products

Product	Nutraceutical Inventory	Health Benefits	References
Chips	Incorporation of essential oils.	It reduces the trans-fat in food diet.	Lee et al, 1986
Soy chilly Sause	Incorporation of natural preservative	It prevents cardiovascular diseases and type 2 diabetes.	Zoinon et al,2017 Kusuma et al,2015
Dhokla	Incorporation of spices.	It helps in lowering body cholesterol level.	Nwofia et al,2013
Biscuit	Incorporation of fish protein concentrate	It acts as a dietary fiber and improves the prebiotic characteristics of the final product.	S.M. Ibrahim et al,2009 Lubna,2012
Instant upma mix	Incorporation of foxtail millet and garden cress seed	It has anti-diabetic, hypo-cholesterolemic, diuretic, anticancer properties.	Rodge et al,2018
Pickles	Incorporation of kale	It has dietary fiber. It helps in proper digestion.	Faik et al;2001
Poha	Incorporation of the combination of rice, wheat and pulses.	It provides instant energy. It has high amount of fibers.	Pai et al;2005

Nutraceutical inventory in salt based product

Sema et al.,(2017) products like spreads are prepared from the salt with the incorporation of carob flour which has various health benefits. It is the rich source of dietary fiber which helps in proper digestion and maintaining the body metabolism. It is very useful in preventing constipation. The product consist of major and minor ingredients. The major ingredients includes the carob flour and hydrogenated palm oil and minor ingredients includes skim milk poder, soya flour,lecithin and hazelnut puree whereas, the minor ingredients were skim milk powder(10%), soyabean flour (5%), lechithin (1%) and hazelnut puree (1%). For the preparation of this spread, all the ingredients were homogenized using Ultra-Turrax T50, IKA Labortechnik, Staufen, Germany) at 9500 rpm for 10 min. The major ingredients, carob flour and hydrogenated palm oil were blended with constant content of 80%. The other minor ingredients were kept constant at ratio of 20%. The five samples were prepared and kept for analysis and the sample which possessed the closer instrumental spreadibilty values were selected for further process. Later the spread were kept at ambient temperature (26 ± 2°C) and all other measurements tests were performed within 1 weeks. The overall acceptability were done on the basis of hedonic scale.Rodge et al., (2018) developed instant upma mix are prepared with the incorporation of garden cress seeds. Upma is consumed as breakfast in many places. Kusuma et al., (2015) reported that other types of salt-based products are soybean which are highly pretentious. Different types of products are prepared by soybean such as, soy flour, soy milk, soy flour, soy tofu etc. Zoinon et al.,(2017) reported that the chips are manufactured by the incorporation of essential oils. It helps in curing cardiovascular diseases and type 2 diabetes. Nwofia et al., (2013) reported that the products like dhokla are prepared with the incorporation of spices which helps in lowering down the body cholesterol level in body. Pai et al., (2005) reported that product such as poha was incorporated with rice, wheat and pulses. It provides instant energy to the body and it consist of high amount of fibers.

Antimicrobial effect of salt

Wijnker et al., (2005) reported that salt has the antimicrobial property which further acts as a preservative for food material. It has been reported that the salt is used as the antimicrobial agent for the filing of sausages casing. By addition of salt in it, can preserve for longer time and it acts as a natural preservative for that particular product. salting is the traditional method for the preservation of casings. It inhibits the growth of microorganism and makes the food product effective for consumption.

Updates on jaggery processing

Shrivastav et al., (2016) reported that the jaggery is used as the natural sweetener which is derived from the sugarcane. The processing takes place on the small scale and it is done by the farmers. The juice is extracted from the sugarcane and further filtration and boiling is done.Iron pan is used for it and continuous stirring is required and then the addition of soda is done at required quantity. The scum needs to be removed to get the clear golden color of jaggery.

After the continuous stirring and cooking the consistency of the juice becomes thick enough and it is poured in the iron pan which should be of small or medium size. After this the product is kept for cooling so for the block formation of jaggery. Generally processing of jaggery includes three basic steps which are, extraction, filtration and cooling. Different forms of jaggery are processed such as, liquid jaggery, granular jaggery, solid jaggery etc.

Processing of liquid jaggery

It is the liquid form of jaggery which is obtained during the concentration of pure sugarcane juice and is of semi-solid consistency. The temperature plays a vital role in formation of product. the quality of final product depends upon the quality of sugarcane juice taken for the preparation of liquid jaggery. The temperature need for the production of this product should be range between 103-106°C. Crystallization needs to be avoided and for that some amount of acid is added.

Processing of granular jaggery

In this processing technique concentrated slurry is added and then rubbed with wooden scrapper for the formation of grains. After that the granular jaggery is sieved and then it is cooled. Crystal formation of size less than 3mm is found to be better in terms of quality parameters. Later the moisture content is reduced up to 2% and the final product is packed in the polyethylene polyster bags.

Processing of solid jaggery

After the filtration the cane juice is then pumped into the open pan and then it was heated continuously. The bagasse fuel is used for this purpose. Later on, the jaggery is concentrated and formed into desired shape and size.

Jaggery based products

There are number of products prepared from the jaggery. Tidke et al., (2017) reported that products such as peanut and chickpea nut brittle are prepared from jaggery. The product was very rich in protein. For the prepation of this product, syrup of sugar, jaggery and corn are prepared in the ratio of 1:1:0:3. Sodium bicarbonate was also added for the preparation of chikki with nuts. Different types of other products such as sweets and sweet snacks are prepared by the addition of jaggery

Nutraceutical inventory in jaggery

Different types of products are prepared from jaggery and incorporation of some other ingredients made it rich nutraceutical aspects. Nath et al.,(2015) reported that the jaggery powder was prepared from the dried form of jaggery by the incorporation of mineral salt ions in it. It reduces the risk of anemia. Jaswant et al., (2013) reported that the payasum is the sweet prepared from the jaggery by the incorporation of bamboo seed. It helps in maintaining the acid balance in the body. Ananthan et al.,(2013) reported that product such as bars were prepared by jaggery with the incorporation of nuts. It helps in reducing the risk of diabetes. Khusboo et al.,(2014) reported that the product such as kheer was produced with the jaggery, along with the incorporation of cereals and skim milk. It provides energy to the body. Khan et al., (2011) reported that the product like chocolate are produced from the jaggery along with the milk powder and coffee powder. It also reduces the risk of osteoporosis.

Antimicrobial effect of jaggery

Shubhra et al., (2009) reported that the jaggery has some antimicrobial property. Microorganism can cause adverse health effect if it contaminates the product. incorporation of jaggery with food product reduces the risk of microorganism and hence it acts as an antimicrobial agent for this product. Different isolates of strains were taken for finding the microbial content in jiggery.

References

Ayaz, F. A., Glew, R. H., Millson, M., Huang, H. S., Chuang, L. T., Sanz, C., & Hayırlıoglu-Ayaz, S. (2006). Nutrient contents of kale (*Brassica oleraceae* L. var. acephala DC.).*Food Chemistry*, *96*(4), 572-579.

Aydın, S., & Özdemir, Y. (2017). Development and characterization of carob flour based functional spread for increasing use as nutritious snack for children. *Journal of Food Quality*, 1-7.

Besbes, S., Drira, L., Blecker, C., Deroanne, C., & Attia, H. (2009). Adding value to hard date (Phoenix dactylifera L.): compositional, functional and sensory characteristics of date jam. *Food Chemistry*, *112*(2), 406-411.

Bolla, K. N. (2015). Soybean consumption and health benefits. *International Journal of Scientific and Technology Research*, *4*(7), 50-3.

Bouchard, D. R., Ross, R., & Janssen, I. (2010). Coffee, tea and their additives: association with BMI and waist circumference. *Obesity Facts*, *3*(6), 345-352.

Chand, K., Singh, A., Verma, A. K., & Lohani, U. C. (2011).Quality evaluation of jaggery chocolate under various storage conditions.*Sugar Tech*, *13*(2), 150-155.

Chandegara, M., Chatterjee, B., & Sewani, N. (2018). Development of novel chocolate energy bar by using nuts. *International Journal of Food and Fermentation Technology*, *8*(1), 93-97.

Durack, E., Alonso-Gomez, M., & Wilkinson, M. G. (2008). Salt: a review of its role in food science and public health. *Current Nutrition & Food Science*, *4*(4), 290-297.

Fechter, W. L., Kitching, S. M., Rajh, M., Reimann, R. H., Ahmed, F. E., Jensen, C. R. C., & Walthew, D. C. (2001). Direct production of white sugar and whitestrap molasses by applying membrane and ion exchange technology in a cane sugar mill.In *Proc. Int. Soc. Sugar Cane Technol* (Vol. 24, pp. 100-107).

Gupta, K., Verma, M., Jain, P., & Jain, M. (2014). Process optimization for producing cowpea added instant kheer mix using response surface methodology. *Journal of Nutrition Health and Food Engineering*, *1*(5), 00030.

Hu, C. H., He, J., Eckert, R., Wu, X. Y., Li, L. N., Tian, Y., ...& Spackman, S. (2011). Development and evaluation of a safe and effective sugar - free herbal lollipop that kills cavity - causing bacteria. *International Journal of Oral Science*, *3*(1), 13.

Ibrahim, S. M. (2009). Evaluation of production and quality of salt-biscuits supplemented with fish protein concentrate. *World Journal of Dairy and Food Sciences*, *4*(1), 28-31.

Ingle, M., Patil, J., & Nawkar, R. (2016).Nutritional evaluation of sugar free aonla candy.*Asian Journal of Dairy & Food Research*, *35*(4).

Katke, S. D., Patil, P. S., & Pandhare, G. R. (2018). Process standardization and development of honey based aonla (Phyllanthus emblica) candy. *Food Science Research Journal*, *9*(2), 311-317.

Kumar, K. S., Bhowmik, D., Biswajit, C., & Chandira, M. R. (2010). Medicinal uses and health benefits of honey: an overview. *J Chem Pharm Res*, *2*(1), 385-395.

Mat Sharif, Z., Mohd Taib, N., Yusof, M. S., Rahim, M. Z., Tobi, M., Latif, A., & Othman, M. S. (2017). A study on shelf life prolonging process of chili soy sauce in Malaysian SMEs'(small medium enterprise). In *IOP Conference Series: Materials Science and Engineering* (Vol. 203, No. 012026, pp. 1-6). IOP Publishing.

Nath, A., Dutta, D., Kumar, P., & Singh, J.P. (2015). Review on recent advances in value addition of jaggery based products. *Journal of Food Processing and technology*, 6(4): 1-4

Padmashree, A., Sharma, G. K., & Govindaraj, T. (2013). Development and evaluation of shelf stability of flaxoat nutty bar in different packaging materials. *Food and Nutrition Sciences*, *4*(05), 538.

Pramanick, P., Zaman, S., & Mitra, A. (2014). Processing of fruits with special reference to S. Apetala fruit jelly preparation. *International Journal of Universal Pharmacy and Bio Sciences*, *3*(5), 36-49.

Radia, A., Boukhiar, A., Kechadi, K., & Benamara, S. (2018). Preparation of a Natural Candy from Date (*Phoenix dactylifera* L.), Olive (*Olea europaea* L.), and Carob (*Ceratonia siliqua* L.)Fruits.*Journal of Food Quality*, pp 1-9

Rodge, S. M., Bornare, D. T., & Babar, K. P. (2018). Formulation and quality evaluation of instant upma mix of foxtail millet and garden cress seed. *International Journal of Chemical Studies*, *6*(3), 1854-1857.

Singh, J., Solomon, S., & Kumar, D. (2013). Manufacturing jaggery, a product of sugarcane, as health food.*Agrotechnology,* S11 (007): 1-3.

Srivastav, P., Verma, A. K., Walia, R., Parveen, R., & Singh, A. K. (2016). Jaggery: A revolution in the field of natural sweeteners. *European Journal of Pharmaceutical and Medical Research*, 198.

Subramanian, R., Umesh Hebbar, H., & Rastogi, N. K. (2007).Processing of honey: a review. *International Journal of Food Properties*, *10*(1), 127-143.

Tidke, B., Sharma, H. K., & Kumar, N. (2017).Development of peanut and chickpea nut brittle (Chikki) from the incorporation of sugar, jaggery and corn syrup.*International Food Research Journal*, *24*(2), 657.

Tong, Q., Zhang, X., Wu, F., Tong, J., Zhang, P., & Zhang, J. (2010).Effect of honey powder on dough rheology and bread quality.*Food Research International*, *43*(9), 2284-2288.

Verma, I., & Gupta, R. K. (2015). Estimation of phytochemical, nutritional, antioxidant and antibacterial activity of dried fruit of sacred figs (Ficus religiosa) and formulation of value added product (Hard Candy). *Journal of Pharmacognosy and Phytochemistry*, *4*(3), 257.

Wijnker, J. J., Koop, G., & Lipman, L. J. A. (2006). Antimicrobial properties of salt (NaCl) used for the preservation of natural casings. *Food Microbiology*, *23*(7), 657-662.

Zumbe, A., Lee, A., & Storey, D. (2001). Polyols in confectionery: the route to sugar-free, reduced sugar and reduced calorie confectionery. *British Journal of Nutrition*, *85*(S1), S31-S45.

4

Rheological and Thermal Changes Occurring During Processing

Renuka Singh, Mamta Bhardwaj, D.C. Saxena

Department of Food Engineering & Technology, Sant Longowal Institute of Engineering & Technology, Sangrur, Punjab

Introduction

Food processing comprise of techniques employed to transform raw ingredients into final food product or to preserve the food product. Food processing industry as well as at domestic scale the food processing methods aim at providing processed food products for daily consumption by humans and animals. Food processing aims at increasing the shelf life, preserve, make nutritious or ready to eat foods, and getting the best quality final products. There are various techniques and methods which come under food processing such as drying, dehydration, fermentation, pickling, freezing, mixing, pumping, pasteurization, cooking and homogenization etc.

Rheological properties of foods are greatly influenced by various unit operations and processing methods or conditions such as heating and mechanical treatments. Processing of different food products such as dairy, fruit and vegetable products like jams and purees, chocolates, dough development etc. and basic make up of foods such as fats, proteins and others such as emulsifiers, hydrocolloids etc. affects the rheological properties of food product and which ultimately affects the quality of end product. Apart from processing conditions the compositional variations have a great impact on the textural or rheological properties of foods. The changes in rheological characteristics of a food product reflects the alterations in the molecular structure of the food material, this can aid in revealing the changes in product structure during different processing methods or conditions. In designing of various processes such as cooling, freezing and equipment design information about thermal properties is crucial. Moreover, in order to design storage structures for foods and refrigeration equipment the knowledge of thermal properties is important (Becker et al. 2003).

For rheological measurements various tests and techniques are available, such as steady and dynamic shear measurements. From the steady shear properties flow properties of the sample can be revealed whether the flow is Newtonian or Non Newtonian. In case of dynamic shear rheological properties information on viscoelastic characteristics can be withdrawn in terms of storage and loss modulus and depending on the predominance of each modulus the nature of sample can be predicted i.e. whether viscoelastic solid or viscoelastic fluid (Bhardwaj et al. 2019). Apart from these measurements, other rheological measurements are available for specific food materials, such as farinograph, alveo-consistograph, mixograph etc. are meant for study of rheological properties of dough. So, accordingly, depending upon the food material choice of rheological measurement can be made. To study the thermal transitions DSC (Differential scanning calorimeter) and DTA (Differential thermal analysis) are commonly employed. Dynamic measurements as a function of temperature or time are made in these techniques to detect the endothermal and exothermal changes. These techniques can be used to determine the phase transitions in various foods, for instance, crystallization and melting of water, lipids and other food components; denaturation of protein; gelatinization of starch and others (Roos 2015).

Effect of processing on fruit and vegetable products

Rheological properties of fruit purees, juices and other fruit based products such nectars, ice creams and jellies which consist of pulp as raw ingredient are vital to study to elucidate the effect of processing on the crucial quality parameters of the final product. The most important rheological characterization in case of juices and purees is the shear-stress against the shear rate relationship. Rheological characterization is also needed for the design of equipment and selecting the processing parameters for processes such as heating, separating and mechanical operations (Yeow et al. 2002, Steffe et al. 1992).

The plant based dispersions as that from fruits and vegetables are composed of insoluble fractions along with aqueous solution. Thus, there are various factors which affect the rheological properties such as solid content, particle size distribution and serum viscosity, these in turn are affected by processes such as heating, cooling and mechanical treatments (Rao et al. 1992). The purees from fruits and vegetables in general are shear thinning in nature and they do exhibit some yield stress (Rao et al. 1977). As per a study of Espinosa et al. (2011) increasing the pulp content increased the shear thinning behavior as well as the consistency coefficient. It was also elucidated from the study that decreasing particle size and pulp content decreased the apparent viscosity of apple puree. Mechanical processing methods such as refining (sieving) High pressure homogenization and others affect the particle size distribution. Thus, varying the screen size will alter the particle size and pulp content which will ultimately affect the rheological properties, so refining operations are to be chosen carefully (Colin-Henrion 2009).

The linear viscoelastic region obtained from dynamic rheological analysis is dependent on the pulp content which varies depending upon the processing conditions. Lower pulp content results in a smaller linear viscoelastic region and vice versa. The purees behave like soft solids. Cooking also affects the rheological properties of puree and serum. In the same study, heat treatment led to decrease in viscosity because of the effect on cell wall structure (Espinosa et al. 2011). Heat treatment leads to degradation and solubilization of pectins thereby altering the rigidity of particles which ultimately affects the rheological properties (Redgwell et al. 2008).

Pasteurization is a common practice in preservation of food materials, so its effect on rheological properties has been studied on juices by various researchers. Generally the flow behavior is altered as in case of white carrot juice as studied by Nadulski et al. (2015). The untreated juice showed Newtonian behavior whereas the pasteurized juice possessed pseudoplastic character. Similarly, other thermal processing techniques such as blanching employed to inactivate enzymes where the fruits and vegetables are scalded in boiling water, bring about changes in rheological properties of plant based suspensions. Thermal processes result in structural modifications in the plant tissue based foods due to depolymerization of pectins, inactivates the pectinolytic enzymes loss of turgor pressure (Moelants et al. 2014). These factors can result in decreased mechanical rigidity. For instance, storage modulus (G') decreased two fold when rocket puree with stem was blanched at 85°C for 3min as reported by Ahmed et al. (2013). The thermal processing conditions also play detrimental role in the rheological properties. Likewise, the time and temperature conditions as reported by Usiak et al. (1995) affected the rheological properties of sauce prepared from blanched apple. The consistency, yield stress and thickness of sauce were dependent on blanching temperature. It was reported that thickness of sauce decreased at blanching temperatures if 35-59 °C whereas it increased for range of 59-71 °C.

Effect of processing on rheological properties of dairy products

In order to increase the shelf life of milk based products and to manufacture commercial dairy products such as cream, butter, yogurt, fermented milks, cheese and powdered milk various processes comes into play which ultimately affects the rheological properties of the ingredients or the final product. Rheological properties are required to be controlled for new product development aimed at incorporating functional and health benefits, such as, low fat and low sugar ice cream, fat mimicking products so as to avoid defects related to body and texture. Milk and cream shows rheological behaviour analogous to emulsions and suspensions, but depending upon the processes and composition these can show Newtonian or non-Newtonian behaviour (McCarthy et al. 2011). Yogurt, a fermented milk, possess time dependent shear thinning behaviour which is greatly dependent on the pre-treatment of milk, starter culture and incubation conditions (Costa et al 2015, Delgado et al.2017). Brighenti et al. (2018) studied the effect of processing conditions of homogenization pressure and fermentation temperature on the rheological and textural properties of acid gel of cream cheese. Cream cheese is formed after an initial gel formation from high-fat milk and after that it involves series of processing steps such as shearing, heating and dewatering. It was revealed that with increase in fermentation temperature (20 to 25°C) and homogenization pressure (10 to 25 MPa) consistency of the acid gel increased. Lower stiffness was observed in cream cheese at higher fermentation temperature and homogenization pressure because of the combined effect of, (i) increment in absorption of protein at the fat globule interface and thus reduction in bulk protein content at higher homogenization pressures, (ii) higher fermentation temperature promoted coarser gel network. In another work by Labropoulos et al. (1984), effect of processing under ultra-high temperature, conventional vat temperature and pasteurization on rheological properties of yogurt was studied. Lower gel firmness and lower apparent viscosity, but higher spread ability and fluidity of yogurt were reported under ultra-high temperature processing as compared to other techniques.

Effect of processing on rheological properties of chocolates

Chocolate is a multi-phase confectionery product which consists of non-fat particles dispersed in continuous phase, cocoa butter. In terms of eating quality and chocolate processing rheological properties are of utmost concern in its molten state. The rheological properties of chocolate are intently related to its microstructure being imparted by raw ingredients and processing conditions. Molten chocolate is a suspension with properties that are strongly affected by particle characteristics including not only the dispersed particles but also the fat crystals formed during chocolate cooling and solidification (Goncalves et al. 2010). Information about viscosity of chocolate is crucial for different applications, for instance, coating of chocolate on candies require an optimum viscosity, if it will be too low then the weight of chocolate over candy will be low and too high will result in bubble formation (Beckett et al.

2009b). Basic processing steps of chocolate consist of mixing, refining, conching, tempering, molding and packing (Attaie et al. 2003). Any alterations in the processing parameters of these can result in change of rheological properties. Chocolate possess non-Newtonian behaviour. Among the mathematical models, Casson model is best suitable for chocolate flow properties (Wolf et al. 2017, Glicerina et al. 2015). Casson model is given by equation 1.

$$\sqrt{\tau} = K_o + K_1\sqrt{\dot{\gamma}} \quad (1)$$

Where, K_0 and K_1 are constants which depend on the viscosity of chocolate, τ is shear stress and $\dot{\gamma}$ is the shear rate. In some cases, behaviour of tempered chocolate was well described by Herschel-Bulkley constitutive model given by equation 2. (Briggs et al. 2004).

$$\tau = \tau_o + K(\dot{\gamma}^n) \quad (2)$$

Where, τ_0 is the yield stress, K is consistency coefficient and n is flow behaviour index.

During tempering the chocolate is subjected to shear, altering the shear rate and shear time conditions result in chocolate having different quality attributes which can be inferred by studying the rheological properties. Briggs et al. (2004) investigated the rheological properties at 15 and 30 s^{-1} shear rate for 0, 400, 600 and 800 seconds. The holding time and shear rate affects the lipid crystallization which ultimately affects the apparent viscosity of chocolate. The magnitude of apparent viscosity at lower shear rate was higher and vice-versa i.e. shear thinning character was observed. Holding time also significantly affects the apparent viscosity of chocolate, for instance, in the same study it was reported that shorter holding time resulted in lower apparent viscosity irrespective of the shear rate. Thus, choice of processing conditions of tempering is of utmost importance. Sometimes chocolate composition has a more profound effect on the rheological properties of chocolate as compared to the processing parameters. Ingredients such as milk powder, emulsifier type, cocoa butter equivalent type etc. affects rheological properties significantly. Ashkezary et al. (2017) reported that type of emulsifier greatly affected the apparent viscosity of compound chocolate whereas refining time had no significant affect. Similarly, in case of milk chocolates the rheological characteristics are influenced by the formulation and processing steps. Glicerina et al. (2014) investigate the effect of each processing step of chocolate manufacturing on the rheological properties. Microstructural changes and particle size reduction take place after every processing step resulting in changes in interaction among the molecules thus bringing out significant changes in rheological parameters.

Effect of processing on rheological properties of meat

In addition to preservation processing of meat is done to add variety to the diet because processing imparts textural and flavour changes in meat. Processing of meat involves various methods such as smoking, freezing, mincing, grinding, chopping, salting and curing, addition of seasonings, various heat treatments. Rheological properties of meat are crucial to study so as to establish quality grade for the final product.

Thermal processing of meat by cooking, smoking etc. cause the meat proteins to denature resulting in structural changes due to cell membrane destruction, coagulation and gel formation from myofibrillar and sarcoplasmic proteins, tearing of muscle fibers. All these factors bring about changes in the toughness, firmness and texture of the meat. These rheological changes are mainly attributed to the changes in proteins. Grujic et al. (2014) reported the effect of cooking and roasting in the temperature range of 51°C to 100 °C on rheological properties of *M. longissimus dorsi* of pork. Increase in temperature resulted in increase in hardness and firmness in both the thermal processes, but samples processed by roasting showed significant effect on these parameters. Choice of processing conditions also depends on the microbiological safety apart from the optimal rheological properties. Since, in the study mentioned, inspite of satisfactory rheological properties below 71°C it is not wise to proceed for this range of heating temperature because of insufficient microbiological safety as per the American Meat Science Association. Other processing treatments, such as, high pressure processing is used to reduce the salt and phosphates in different meat products such as frankfurter-type sausages, pressure treatment also alters the gel forming capability of proteins thereby affecting the textural properties of processed meat. Sometimes processing involves multiple processing steps, and the rheological properties are affected differently in individual and combined treatments. Fernandez et al. (1997) studied the effect of pressure/thermal treatment on the rheological properties of meat batters. The study revealed that thermal processing alone was better in forming a quality product whereas pressure/thermal treatment impaired the gelling properties of meat thus resulting in undesirable pressurized batters.

Thermal properties and processing effects

Different thermal processing techniques such as pasteurization, sterilization, UHT, balancing and retort have been largely used for food processing industry to maintain undesirable enzymes, microorganisms, and bacteria in food. To have a stable and long shelf life with desired food safety different altered forms of heat treatment processes are being used which include long time heating at high temperature, use of physical and chemical reactions in foods (Saxena et al. 1995). Changes in storage conditions lead to quality degradation such as changes in color, off-odor development, and degradation of freshness and nutrients both during and after thermal processing for different food groups. Every food industry has a series of defined food processing steps that most of the time consists of specialized thermal process that leads to changes in thermal properties of food groups.

A food group can be defined as the collection of different foods that share common or similar nutritional and biological class (Verma et al. 2018). Nutritional compass basically describes food into different food groups and also defines daily serving of group for nutritious diet. USDA has described food in different 5 groups Dairy, Fruits, Grains, Meat and Vegetables.

Thermal properties in food groups are important to understand relations between food quality and processing as change in temperatures during thermal process may lead to both loss of nutrients and shelf life. Food groups are integrated mixtures of proteins, carbohydrates, lipids and water. All thermal properties are equally important to understand relation between processing and quality of food. Hence, different thermal analytical processes such as differential scanning calorimetry (DSC), Dielectric analysis (DEA) and dynamic mechanical analysis (DMA) being used commonly for physiochemical characterization of food groups. The requirement of these thermal analytical techniques for analyzing physiochemical properties of food groups and foods has become very important as it was observed that the physiochemical properties in frozen, low moisture and concentrated systems are linked to non-equilibrium state of the food. To obtain sufficient information for controlling quality changes during food processing and storage, thermal analytical (Fellows et al. 2009) techniques are very useful.

Thermal properties of food groups traditionally change with changes in phase transitions of food that are usually defined according to thermodynamics happening at transition temperatures. Many concentrated food groups i.e. low in moisture and frozen in state representing equilibrium state tends to form non-crystalline and amorphous structures (Spreer et al. 2017). Glass-transition changes in such food groups alters stability of food such as lumpiness and stickiness in powders, crunchiness of cereals breakfast and snacked food, sugar crystallization, frozen food recrystallization, ice formation, and enzymatic reactions in stored foods (Venir et al. 2007).

Linkage between different forms of foods during processing such as glass transition, plasticization and crystallization can be identified by different thermal analytical techniques in different food groups. Thermal properties of each food group vary from other due to their unique structural mixture of lipids, proteins, water, fats, and carbohydrate. These unique mixtures of components in food groups are responsible for different glass-transitions (amorphous, non-crystalline structures), phase transition (melting, crystallization), and state transition (denaturation, gelatinization) that are leading to several changes in state of food groups. (Lopez et al. 2009). Hence, different structures such as amorphous or partial amorphous, and crystalline in foods are formed during food processing due to changes in thermal properties resulting addition or removal of water. Furthermore, basis on the rate of removal or cooling of solvent into amorphous, crystalline or solid different thermal properties can be observed.

Dairy processing

In development of dairy industry different classical milk product classes and technologies are developed: butter, yoghurt, fluid milk, cheese, and different long shelf life products which are results of different milk processing techniques. With the increase in innovations and renovation in dairy technology varieties of technology machinery, and processed have been (Lopez et al. 2007) developed.

Effect of freeze-drying on thermal properties of yoghurt

In freeze drying for dairy products a process called Lyophilisation used to freeze the product, where by using high-pressure vacuum water is extracted in vapour form. The vapour is than collected below freezing chamber on condenser (Truong et al. 2014). Than a sudden rise in temperature extracts all bound moisture from product to preserve it for storage or transportation. In a study done by Elena Venir to develop a stabilized and calcium fortified yoghurt for consumption in space, both sucrose enriched and with added extracts of blueberries were subjected to freeze-drying.

As it is known that stability of product for freeze drying can be improved by increase in glass transition *Tg* and (Walstra et al. 1993) initial melting *Tm* temperatures by changing product formulations that was done during this study. Appropriate parameters of freeze-drying were derived from thermal analysis of yoghurts, considering the different appropriate amounts of added ingredients. DSC traces method was used for analyzing these both thermal properties by formation of sample at a solid concentration of 30% (w/w) (Itoh et al. 1976). Results of DSC traces showed that sucrose addition increases both *Tg* and *Tm,* while these parameters did not show any change where blueberry extracts were added (Maroulis et al. 2002). Increase in glass transition and initial melting temperatures by changing formulation is observed which increases product stability for freeze drying without affecting rheological properties of yoghurt. Whereas in drying step of the process structural bonding of (Supavititpatana et al. 2007) product get weakened because of mechanical energy that is required for reconstitution of water. However, the viscoelastic properties of both samples were retained by formulating amount of water. The sucrose addition in formulation of yoghurt with or without extracts of blueberries increases the stability of yoghurt (Ahmed et al. 2007) by reducing mortality rate of lactic acid bacteria (LAB).

Effect of heating on thermal properties of milk fat

A study on thermal properties of milk was carried by Lopez et al. (2009) to determine effect of heating on both thermal and crystallographic properties. Use of DSC and synchrotron X-ray diffraction (XRDT) was employed for analyzing milk fat and its fractions (Baik et al. 1999) The uses of XRDT and DSC methods facilitates the characterization of both milk fat and its fractions and also relates structural changes with thermal events.. Heating effect on milk fat determined by analyzing melting properties of milk fat, stearin fraction and olein fraction on heating after cooling. Stearin fraction is rich in long chain saturated fatty acid (majorly palmitic, stearic and myristic acid) whereas Olein fraction is rich in TG and on short chain fatty acid (butyric acid) or one unsaturated fatty acid (olein acid). Heating properties were analyzed by following a protocol in which heating of milk for 5 min at 70°C and then cooling from 70°C to -7 °C, and again heating from -7 °C to 60°C was done (Baik et al. 2001). The coupling of DSC and XRDT during melting of milk fat, olein and stearin fractions during heating-cooling-heating at different angles setting of DSC permits the determination of liquid and solid phase percentage. Whereas XRDT permits the quantification of each solid phase as a function of temperature on heating of olein and stearin fraction (Marcotte et al. 2008) By differential scanning calorimetry with coupling of synchrotron X-ray diffraction showed that during heating of olein and stearin recrystallization occurred with the formation of solid-liquid phase (Carson et al. 2006).

During different Dairy technology processing both supramolecular and surface composition of fat gets affected at a large scale. In these dairy processes fat may also be present in continuous phase same as in butter, while in manufacturing and ripening of cheese (pressing of curd globules) MFGM disruption and free fat formation takes place. Furthermore, contributes to lipolytic enzyme activity for fat lipolysis by formation of whey pockets in the surrounding of fat which facilitates localization of fat-protein interface after MFGM disruption (carson et al. 2005). Determination of content of solid fats in dairy products can be done by using Differential scanning Calorimeter (DSC). Increased in the unsaturated fatty acid amount in milk fat to enhance its nutritious properties alters in solid fat content.

There is an exponential relationship observed in general between the droplet size and crystallization temperature for dairy based emulsions, which indicates that smaller the droplet size that ranges below 0.5 um, higher the magnitude in decrease of crystalline temperature (Carson et al. 2006). Emulsifiers are well known for their influencing roles in crystalline behavior of fats for dispersed systems. In emulsified droplets nucleation get initiates by acting of hydrophobic emulsifier tail group as a template. Size distribution of the droplets describes stability of dairy products such as smaller droplets are always more stable against creaming, flocculation and coalescence. But, it is (Gonzo et al. 2002) always difficult to make droplets smaller enough to support product stability that means droplet break-up is more essential than droplet making for dairy products stability. Protein milk denaturation is observed using differential thermal analysis. Solution of protein produces an endothermic peak of characteristic shape and temperature. Alpha lacto albumin has (Zhang et al. 2007) week endothermic peak compare to beta lacto albumin different peaks indicate the different state like the solid proteins have bimodal exothermic peaks between 200 and 400 c which implies oxidative degradation and determination of coagulation of protein.

Effect of processing on thermal characteristics of meat

Animal flesh which is eaten as food is considered as meat animal flesh like sheep chickens, rabbits, pigs and cattle. Main composition of meat is protein (myofibrillar protein actin and myosin), water, and fat. DSC thermo grams of unprocessed meat and processed meat were observed in which denaturation temperature of actin and myosin peak shift to higher temperature (Martens et al. 1982). Thermal behavior of proteins is dependent on different factors like fat. The large amount of fat in meat influences the thermal transition (Kijowski et al. 1988).

Characterization of thermo-physical properties of meat and poultry between cookers or smoke house is very important to study their influence on thermal exchanges leading to changes in meat/poultry emulsions, as both cooking and cooling relies on heat conduction between emulsions (Quinn et al. 1980). To measure/observe thermal conductivity of meat/poultry products thermo-physical properties, different models can be used such as differential scanning calorimeter (DSC), guarded hot plate method, line heat source probe method and capped column test device (Trout et al. 1983).

In a study done on thermo-physical behavior of meat and poultry emulsions to evaluate thermal conductivities shows that there is a linear increase in thermal conductivities with increase in temperature between 20 to 60°C. Above 60°C up to 80°C, there is no change in thermal conductivity of meat and poultry products except bologna. Also there is slight decrease in densities observed during this study at temperatures from 20 to 40°C while a transition phase observed at 40 to 60°C and further decreased from increase in temperature from 60 to 80°C, since there is a decrease of 50kg.m^{-3} in density observed for a raw meat product at room temperature and when product heated at 80°C due to setting or gelation of the meat/poultry product structures (Quinn et al. 1980; Trout et al. 1983). Both densities and thermal conductivity of meat/poultry products are influenced by carbohydrate content (i.e. increase in carbohydrate content results decrease in density). Also the salt content in meat/poultry products affects both thermal conductivities and diffusivity, sometimes these parameters can get increase due to increase in moisture (Kijowski et al.., 1988). Heating of meats/poultry products always results in change of flavors, appearance, nutritional values, and texture. Most of these changes such as tissue shrinkage, discoloration, toughing of muscle-tissues and release of fluid during heating are due to changes in muscle proteins (Byrne et al. 2002).

To study thermal behavior of muscle proteins in different meats differential scanning calorimetry (DSC) can be used. Major transformations in breast muscle's myofibrils washed with water displayed at 55 deg. and 78 deg. respectively corresponding to myosin and actin (Byrne et al. 2001). In poultry meat processing industries usage of pyrophosphate and tripolyphosphate has increased due to their tendency to enhance water binding and to lessen cooking losses, increase in phosphates usages also linked to consumer due to widely increased demands for low sodium level in foods (Astruc et al. 2007). Presence of NaCl in high concentrations destabilizes heat resistance stability due to reduction in denaturation temperature of the protein in meat specimen and myofibril (Davies et al. 2001). While the presence of pyrophosphate (PP) and tripolyphosphate (TPP) in 0.25% and 0.50 % concentration respectively increase myosin thermal stability. From different studies it can be concluded that presence of fat or mechanical processing of meat does not affect protein structural stability, however presence of NaCl affects protein structural stability by reducing (Promeyrat et al. 2010) denaturation temperature. In cooking process for meat products generation of free radicals take place in high amount which results in both proteins and lipids oxidation. Oxidation in lipids produce off flavor and odor in meat product (Athmaselvi et al. 2014) while protein oxidation leads to aggregation of protein in meat due to cross-linking in free amino acids and carbonyl group, which are produced as result of basic amino acids oxidation during cooking. To avoid oxidation or aggregation of protein and off flavor during cooking, addition of antioxidants in meat products during processing can limit the lipid and protein oxidation up to significant amount. However, to avoid any thermal denaturation in meat products during processing long duration and high temperature cooking must be avoided (Maggio et al. 2012).

Effects of processing on thermal properties of fruits

For extending the shelf-life and to preserve physical, chemical and sensory properties of fruits and vegetables, relative humidity and temperatures are very critical factors. Thus, dehydration process has been considered an important and efficient alternative for fruit storage because it results in declining water activity which suppresses both chemical and enzymatic reactions leading to food deterioration (Vasquez caicedo et al. 2007). Different methods have been used for preservation of fruits and vegetables such as hot air drying and osmodehydration (freeze, drying). According to International Confederation for Thermal analysis and Calorimetry (ICTAC), Thermal analysis (Sitte et al. 1980) is defined as a collection of different techniques that studies the relation between food group properties and its temperature. Thermal analysis also allow us to determine the duration and temperature at which drying will complete without degrading food properties. DSC is used to evaluate vegetable oil's quality parameters and oxidative deterioration of oil. Whereas, (Sila et al. 2008) thermogravimetric analysis also used to analyze Okra fiber decomposition during dying which shows three steps of weight loss in two stages.

Fruits that grow in subtropical and tropical regions such as India and many other countries of world are very important crop from economical pint of view are more critical in terms of storage, sensory and shelf life. Both temperature and relative humidity are most critical factors for texture, sensory, shelf life extension, physical and chemical properties preservation of fruits (Miller et al. 2012). Thus, dehydration, sterilization and other thermal treatments could be beneficial for shelf life extension by controlling enzymatic and chemical reactions, water activity that are responsible for food deterioration. Sterilization and dehydration by hot air drying are most commonly used industrial process (Chen j et al. 2014). In both processes alteration in enzymatic reactions and water activity results shelf life extension, traces of which can be seen by using TGA-DSC and IR which shows particle activities with electric and dielectric properties and temperature (He Z et al. 2016).

There is clear relationship between temperature and trans-cis-isomerization of beta-carotene in Mango Puree in pasteurization process, in which mango carotenoids gets naturally dissolve in lipid droplet of globular chromosomes. Continuous heating of these dissolved beta- carotene results degradation of al trans-beta-carotene when heated at 150 deg. for 30 min, while heating at 50-100 deg. for 30 minute is favorable amount for pasteurization where only slightly trans-beta-carotene and cis-isomers gets affected. Whereas, in case of carrot juice beta-carotene degradation observed only after long duration exposure to high temperature continuously. During sterilization of carrot juices for prolonged duration of 40 min at 130°C (Hurtado et al. 2015) higher beta-carotene stability is observed. Reason for high-beta-carotene stability even at longer duration exposure to temperature in carrot juices is due to pigment crystals found in carrot roots. Crystalline beta-carotene in both mango puree and carrot juices is stable at pasteurization temperatures (Majumdar et al. 2011).

However, in physical state change during air-drying of different fruits and vegetables such as butternut, squash, sweet potato and yellow corn at a temperature range of 60-80 deg. For 50 hr. is best observed. Hence, due to variation in thermal stability of carotenoids in different fruits and vegetables specialized treatments for food processing required for each and every food (Zhang Y et al. 2010). In view to analyze texture changes of processed fruits and vegetables observed during food processing due to preheating, high-pressure and high temperature treatments stability of both micro and macro molecular organization crucial to maintain specially pectin. Vegetables that are pretreated with high pressure show significant texture stability during thermal sterilization (Roos Y H et al. 1995).

Different thermal techniques are used in processing of fruit juices and beverages whereas High temperature-Long time in which temperature is higher than 80 deg. And holding time is more than 30 sec. is the most commonly adapted method. Furthermore, it is classified as sterilization (temperature higher than 100 deg.), canning (temperature 100 deg.), and pasteurization (temperature less than 100 deg.). High temperature exposure (string stress) may increase permeability of membrane due to lipid phase transition and protein conformation that (Roberts et al. 2003) are leading to cell death. Changes in Membrane fluidity of fruits differs accordingly with the type thermal stresses. Fruits which are slightly acidic in nature i.e. having pH > 4.5, need high temperature treatments to meet desired shelf life. In case of Aonla, bottle gourd, ginger and lemon during thermal processing minimum and maximum loss of ascorbic acid of juice blend were 22.97 % at 80 deg for 5 min and 47.70 % at 95°C (Lopez et al. 2006).

In apple more aromatic compounds observed in thermal treated juice at 80 deg for 30 min than untreated juice. In smoothie made up of apple, banana, orange, strawberry thermal treatment at 85°C for 7 min benefited regarding enzyme activation (POD, PPO, PME), and resulted in cooked fruit flavor. In juice blend of Basil and bottle gourd thermal treatment at 95 deg. for 15 min results in shelf life of 6 month at room temperature and also observed microbiologically safe (Singh et al. 2007). In Blackberry treating juice at 80-90 deg for 300 min results in reduction of antioxidant activity. However, the amount of cyaniding derivative slightly increased.

References

Ahmed, J. and Ramaswamy, H.S., 2007. Dynamic rheology and thermal transitions in meat-based strained baby foods. *Journal of Food Engineering, 78*(4), pp.1274-1284.

Ahmed, J., Al-Salman, F. and Almusallam, A.S., 2013. Effect of blanching on thermal color degradation kinetics and rheological behavior of rocket (*Eruca sativa*) puree. *Journal of Food Engineering, 119*(3), pp.660-667.

Ashkezary, M.R., Yeganehzad, S., Vatankhah, H., Todaro, A. and Maghsoudlou, Y., 2017. Effects of different emulsifiers and refining time on rheological and textural characteristics of compound chocolate. *Italian Journal of Food Science, 30*(1).

Astruc, T., Marinova, P., Labas, R., Gatellier, P. and Santé-Lhoutellier, V., 2007. Detection and localization of oxidized proteins in muscle cells by fluorescence microscopy. *Journal of Agricultural and Food Chemistry, 55*(23), pp.9554-9558.

Athmaselvi, K.A., Kumar, C., Balasubramanian, M. and Roy, I., 2014. Thermal, structural, and physical properties of freeze dried tropical fruit powder. *Journal of Food Processing, 2014.*

Attaie, H., Breitschuh, B., Braun, P. and Windhab, E.J., 2003. The functionality of milk powder and its relationship to chocolate mass processing, in particular the effect of milk powder manufacturing and composition on the physical properties of chocolate masses. *International Journal of Food science & Technology, 38*(3), pp.325-335.

Baik, O.D., Marcotte, M., Sablani, S.S. and Castaigne, F., 2001. Thermal and physical properties of bakery products. *Critical Reviews in Food Science and Nutrition, 41*(5), pp.321-352.

Baik, O.D., Sablani, S.S., Marcotte, M. and Castaigne, F., 1999. Modeling the thermal properties of a cup cake during baking. *Journal of Food Science, 64*(2), pp.295-299.

Becker, B. R., and B. A. Fricke. "freezing| Principles." (2003): 2706-2711.

Beckett, S. T. The science of chocolate. 2 ed. Cambridge: Royal Society of Chemistry Paperbacks, 2009b. 240 p.

Bhardwaj, M., Sandhu, K.S. and Saxena, D.C., 2019. Experimental and modeling studies of the flow, dynamic and creep recovery properties of pearl millet starch as affected by concentration and cultivar type. *International journal of biological macromolecules, 135*, pp.544-552.

Briggs, J.L. and Wang, T., 2004. Influence of shearing and time on the rheological properties of milk chocolate during tempering. *Journal of the American Oil Chemists' Society, 81*(2), pp.117-121.

Brighenti, M., Govindasamy-Lucey, S., Jaeggi, J.J., Johnson, M.E. and Lucey, J.A., 2018. Effects of processing conditions on the texture and rheological properties of model acid gels and cream cheese. *Journal of dairy science, 101*(8), pp.6762-6775.

Byrne, D.V., Bredie, W.L.P., Bak, L.S., Bertelsen, G., Martens, H. and Martens, M., 2001. Sensory and chemical analysis of cooked porcine meat patties in relation to warmed-over flavour and pre-slaughter stress. *Meat Science, 59*(3), pp.229-249.

Byrne, D.V., Bredie, W.L.P., Mottram, D.S. and Martens, M., 2002. Sensory and chemical investigations on the effect of oven cooking on warmed-over flavour development in chicken meat. *Meat Science, 61*(2), pp.127-139

Carson, J.K., 2006. Review of effective thermal conductivity models for foods. *International Journal of Refrigeration, 29*(6), pp.958-967.

Carson, J.K., Lovatt, S.J., Tanner, D.J. and Cleland, A.C., 2005. Thermal conductivity bounds for isotropic, porous materials. *International Journal of Heat and Mass Transfer, 48*(11), pp.2150-2158.

Carson, J.K., Lovatt, S.J., Tanner, D.J. and Cleland, A.C., 2006. Predicting the effective thermal conductivity of unfrozen, porous foods. *Journal of Food Engineering, 75*(3), pp.297-307.

Chen, J., Tao, X.Y., Sun, A.D., Wang, Y., Liao, X.J., Li, L.N. and Zhang, S., 2014. Influence of pulsed electric field and thermal treatments on the quality of blueberry juice. *International journal of food properties, 17*(7), pp.1419-1427.

Colin-Henrion, M., Mehinagic, E., Patron, C. and Jourjon, F., 2009. Instrumental and sensory characterisation of industrially processed applesauces. *Journal of the Science of Food and Agriculture, 89*(9), pp.1508-1518.

Costa, M.P., Frasao, B.S., Silva, A.C.O., Freitas, M.Q., Franco, R.M. and Conte-Junior, C.A., 2015. Cupuassu (Theobroma grandiflorum) pulp, probiotic, and prebiotic: Influence on color, apparent viscosity, and texture of goat milk yogurts. *Journal of Dairy Science, 98*(9), pp.5995-6003.

Davies, K.J., 2001. Degradation of oxidized proteins by the 20S proteasome. *Biochimie, 83*(3-4), pp.301-310.

Delgado, K.F., da Silva Frasao, B., da Costa, M.P. and Junior, C.A.C., 2017. Differen t alternatives to improve rheological and textural characteristics of fermente d goat products: A review. *Rheol: Open Access, 1*(106), p.2.

Espinosa, L., To, N., Symoneaux, R., Renard, C.M., Biau, N. and Cuvelier, G., 2011. Effect of processing on rheological, structural and sensory properties of apple puree. *Procedia Food Science, 1*, pp.513-520.

Fellows, P.J., 2009. *Food processing Technology: Principles and Practice*. Elsevier.

Fernández-Martín, F., Fernández, P., Carballo, J. and Jiménez Colmenero, F., 1997. Pressure/heat combinations on pork meat batters: protein thermal behavior and product rheological properties. *Journal of Agricultural and Food Chemistry, 45*(11), pp.4440-4445.

Gajera, R.R. and Joshi, D.C., 2014. Processing and storage stability of bottle gourd (L. siceraria). *Agricultural Engineering International: CIGR Journal, 16*(2), pp.103-107.

Gonzalez, M.E. and Barrett, D.M., 2010. Thermal, high pressure, and electric field processing effects on plant cell membrane integrity and relevance to fruit and vegetable quality. *Journal of Food Science, 75*(7), pp.R121-R130.

Gonzo, E.E., 2002. Estimating correlations for the effective thermal conductivity of granular materials. *Chemical Engineering Journal, 90*(3), pp.299-302.

He, Z., Tao, Y., Zeng, M., Zhang, S., Tao, G., Qin, F. and Chen, J., 2016. High pressure homogenization processing, thermal treatment and milk matrix affect *in vitro* bioaccessibility of phenolics in apple, grape and orange juice to different extents. *Food Chemistry, 200*, pp.107-116.

Hurtado, A., Picouet, P., Jofré, A., Guàrdia, M.D., Ros, J.M. and Bañón, S., 2015. Application of high pressure processing for obtaining "fresh-like" fruit smoothies. *Food and Bioprocess Technology, 8*(12), pp.2470-2482.

Itoh, T., Wada, Y. and Nakanishi, T., 1976. Differential thermal analysis of milk proteins. *Agricultural and Biological Chemistry, 40*(6), pp.1083-1086.

Kijowski, J.M. and Mast, M.G., 1988. Thermal properties of proteins in chicken broiler tissues. *Journal of Food Science, 53*(2), pp.363-366.

Labropoulos, A.E., Collins, W.F. and Stone, W.K., 1984. Effects of ultra-high temperature and vat processes on heat-induced rheological properties of yogurt. *Journal of Dairy Science, 67*(2), pp.405-409.

Lopez, C. and Briard-Bion, V., 2007. The composition, supramolecular organisation and thermal properties of milk fat: a new challenge for the quality of food products. *Le Lait, 87*(4-5), pp.317-336.

Lopez, C. and Ollivon, M., 2009. Triglycerides obtained by dry fractionation of milk fat: 2. Thermal properties and polymorphic evolutions on heating. *Chemistry and Physics of Lipids, 159*(1), pp.1-12.

Lopez, C., Bourgaux, C., Lesieur, P., Riaublanc, A. and Ollivon, M., 2006. Milk fat and primary fractions obtained by dry fractionation: 1. Chemical composition and crystallisation properties. *Chemistry and Physics of Lipids, 144*(1), pp.17-33.

Maggio, R.M., Cerretani, L., Barnaba, C. and Chiavaro, E., 2012. Application of differential scanning calorimetry-chemometric coupled procedure to the evaluation of thermo-oxidation on extra virgin olive oil. *Food Biophysics, 7*(2), pp.114-123.

Majumdar, T.K., Wadikar, D.D., Vasudish, C.R., Premavalli, K.S. and Bawa, A.S., 2011. Effect of storage on physico-chemical, microbiological and sensory quality of bottlegourd-basil leaves juice. *American Journal of Food Technology, 6*(3), pp.226-234.

Marcotte, M., Taherian, A.R. and Karimi, Y., 2008. Thermophysical properties of processed meat and poultry products. *Journal of Food Engineering, 88*(3), pp.315-322.

Maroulis, Z.B., Krokida, M.K. and Rahman, M.S., 2002. A structural generic model to predict the effective thermal conductivity of fruits and vegetables during drying. *Journal of Food Engineering, 52*(1), pp.47-52.

Martens, H., Stabursvik, E. and Martens, M., 1982. Texture and colour changes in meat during cooking related to thermal denaturation of muscle proteins 1. *Journal of Texture studies, 13*(3), pp.291-309.

McCarthy, O.J., 2011. Rheology of liquid and semi-solid milk products. *by Fuquay JW, Fox PF, McSweeney PLH (eds.) Encyclopedia of Dairy Sciences, 4*, pp.520-531.

Miller, F.A. and Silva, C.L., 2012. 15 Thermal Treatment Effects in Fruit Juices. *Advances in Fruit Processing Technologies*, p.363.

Moelants, K.R., Cardinaels, R., Van Buggenhout, S., Van Loey, A.M., Moldenaers, P. and Hendrickx, M.E., 2014. A Review on the Relationships between Processing, Food Structure, and Rheological Properties of Plant-Tissue-Based Food Suspensions. *Comprehensive Reviews in Food Science and Food Safety, 13*(3), pp.241-260.

Ollivon, M., Keller, G., Bourgaux, C., Kalnin, D., Villeneuve, P. and Lesieur, P., 2006. DSC and high resolution X-ray diffraction coupling. *Journal of Thermal Analysis and Calorimetry, 85*(1), pp.219-224.

Promeyrat, A., Gatellier, P., Lebret, B., Kajak-Siemaszko, K., Aubry, L. and Santé-Lhoutellier, V., 2010. Evaluation of protein aggregation in cooked meat. *Food Chemistry, 121*(2), pp.412-417.

Quinn, J.R., Raymond, D.P. and Harwalkar, V.R., 1980. Differential scanning calorimetry of meat proteins as affected by processing treatment. *Journal of Food Science, 45*(5), pp.1146-1149.

Rao, M.A., 1977. Rheology of liquid foods-a review 1. *Journal of Texture Studies, 8*(2), pp.135-168.

Rao, M.A., 1992. The structural approach to rheology of plant food dispersions. *Revista Espanola de Ciencia y Tecnologia de Alimentos (Spain).*

Redgwell, R.J., Curti, D. and Gehin-Delval, C., 2008. Role of pectic polysaccharides in structural integrity of apple cell wall material. *European Food Research and Technology, 227*(4), pp.1025-1033.

Roberts, D.D., Pollien, P. and Watzke, B., 2003. Experimental and modeling studies showing the effect of lipid type and level on flavor release from milk-based liquid emulsions. *Journal of Agricultural and Food Chemistry, 51*(1), pp.189-195.

Roos, Y. H. (1995). Food components and polymers. In S. L. Taylor (Ed.), Phase transition in foods (pp. 109–156). London, UK: Academic Press Ltd.

Roos, Y.H. and Drusch, S., 2015. *Phase Transitions in Foods.* Academic Press, 49-77

Saxena, D.C., Rao, P.H. and Rao, K.R., 1995. Analysis of modes of heat transfer in tandoor oven. *Journal of food engineering, 26*(2), pp.209-217.

Sila, D.N., Duvetter, T., De Roeck, A., Verlent, I., Smout, C., Moates, G.K., Hills, B.P., Waldron, K.K., Hendrickx, M. and Van Loey, A., 2008. Texture changes of processed fruits and vegetables: potential use of high-pressure processing. *Trends in Food Science & Technology, 19*(6), pp.309-319.

Singh, J., Kaur, L. and McCarthy, O.J., 2007. Factors influencing the physico-chemical, morphological, thermal and rheological properties of some chemically modified starches for food applications—A review. *Food Hydrocolloids, 21*(1), pp.1-22.

Sitte, P., Falk, H. and Liedvogel, B., 1980. Chromoplasts. *Pigments in Plants, 2*, pp. 117-130

Spreer, E. (2017). Milk and dairy product technology. Routledge.

Steffe, J.F. *Rheological methods in foods process engineering*, 2nd, 428 East Lansing, Mich: Freeman

Supavititpatana, T. and Apichartsrangkoon, A., 2007. Combination effects of ultra-high pressure and temperature on the physical and thermal properties of ostrich meat sausage (yor). *Meat Science, 76*(3), pp.555-560.

Trout, G.R. an Schmidt, G.R. 1983. Utilization of phosphates in meat products. Reciprocal Meat Conference of the American Meat Science Association proceedings, Fargo, ND, Vol. 36, p.24

Truong, T., Bansal, N., Sharma, R., Palmer, M. and Bhandari, B., 2014. Effects of emulsion droplet sizes on the crystallisation of milk fat. *Food chemistry, 145*, pp.725-735.

Usiak, A.G., Bourne, M.C. and Rao, M.A., 1995. Blanch temperature/time effects on rheological properties of applesauce. *Journal of Food Science, 60*(6), pp.1289-1291.

Vásquez-Caicedo, A.L., Schilling, S., Carle, R. and Neidhart, S., 2007. Effects of thermal processing and fruit matrix on β-carotene stability and enzyme inactivation during transformation of mangoes into purée and nectar. *Food Chemistry, 102*(4), pp.1172-1186.

Venir, E., Del Torre, M., Stecchini, M.L., Maltini, E. and Di Nardo, P., 2007. Preparation of freeze-dried yoghurt as a space food. *Journal of Food Engineering, 80*(2), pp.402-407.

Verma, R., Jan, S., Rani, S., Jan, K., Swer, T.L., Prakash, K.S., Dar, M.Z. and Bashir, K., 2018. Physicochemical and functional properties of gamma irradiated buckwheat and potato starch. *Radiation Physics and Chemistry, 144*, pp.37-42.

Walstra, P., 1993. Principles of emulsion formation. *Chemical Engineering Science, 48*(2), pp.333-349.

Wolf, B., 2017. Chocolate flow properties. *Beckett's Industrial Chocolate Manufacture and Use,*, pp.274-297.

Yeow, Y.L., Perona, P. and Leong, Y.K., 2002. A reliable method of extracting the rheological properties of fruit purees from flow loop data. *Journal of Food Science, 67*(4), pp.1407-1411.

Zhang, L., Lyng, J.G. and Brunton, N.P., 2007. The effect of fat, water and salt on the thermal and dielectric properties of meat batter and its temperature following microwave or radio frequency heating. *Journal of Food Engineering, 80*(1), pp.142-151.

Zhang, Y.I., Gao, B.E.I., Zhang, M., Shi, J. and Xu, Y., 2010. Pulsed electric field processing effects on physicochemical properties, flavor compounds and microorganisms of longan juice. *Journal of Food Processing and Preservation, 34*(6), pp.1121-1138.

5

Technological Advancements in Processing of Legumes and Pulses

Deep N.Yadav, Surya Tushir, P.N. Guru, D.K. Yadav and R.K. Vishwakarma*

Indian Council of Agricultural Research-Central Institute of Post- Harvest Engineering and Technology, Ludhiana, Punjab

India is the largest producer of pulses and contributes around 25% of total global pulse production. In addition, it is also world's biggest consumer (27% of world consumption) and importer (14%) of pulses in the world. Pulses account for around 20 per cent of the area under food grains and contribute around 7-10 per cent of the total food grains production in India. Though, pulses are grown in both *Kharif* and *Rabi* seasons, *Rabi* pulses contribute more than 60 per cent of the total production. Gram is the most dominant pulse having a share of around 40 per cent in the total production followed by *tur/arhar* at 15 to 20 per cent and *urad* and *moong* at around 8-10 per cent each. Madhya Pradesh, Maharashtra, Rajasthan, Uttar Pradesh and Karnataka are the top five pulses producing states in India. Considering the nutritional importance of pulses and its products in Indian diet, the production and availability are not sufficient. Since independence, grain production has grown dramatically but pulses production and its availability have declined. On the other hand, the per capita availability of cereal and millets has increased in spite of a four-fold increase in population. The consumption ratio of cereal-to-pulse has risen from 6:1 to 12:1.

Pulse processing is the major need of developing nations as it is the most acceptable and cheap source of protein and hence can help combating Protein Energy Malnutrition (PEM). In addition, the identification and recognition of various newer and improved processing methods can provide the basis for further development of legume processing industry. Traditionally, home cooking is applied for consumption of pulses since ages. Modernization has no doubt created a lot of ease and developed various techniques to make different edible forms of legumes/pulses. Many advancements have been happened in traditional processing methods (milling, fermentation, roasting, boiling, steaming etc). and resulted into new products. Improved techniques for efficient milling of legumes into *dhal* (dehusked splits) are an important area of investigation. A systematic method that eliminates wastes and losses has great significance in the Indian dietary pattern. New systems for processing legumes come under several main categories, namely, improved decortications and milling techniques; quick-cooking pulses; canned pulses; pulse flakes, expanded snack foods etc. Such new and innovative processing technologies presented in this chapter can be applied and used in developing nations. Present chapter will discuss both traditional and non-traditional means of pulse processing.

Nutritional composition

Pulses are crops of *Leguminosae* family whose seed is used for edible purpose and is located within the pod. All the pulses are considered legumes but not all the legumes are pulses. Pulses are regarded as a great crop which contains balanced nutritional composition. They are rich source of protein and are among significant sources of cheap and readily available carbohydrate, protein, dietary fiber, minerals, and vitamins in food (Perez-Hidalgo et al., 1997; Sathe, 2002; Kutos et al., 2003; Osorio-Diaz et al., 2003; Wand and Toews, 2011). Plant proteins are now being regarded as excellent, versatile, and available sources of functional and biologically active food components. Aside from other components, pulses have been, identified as excellent sources of plant proteins (FAO, 2007). Even though, diversity of legumes are eaten in their immature state, the greatest interest from a nutritional point of view is in the consumption of the matured dried grains. Pulses provide around 20 - 30 g of protein per 100 g, twice as much as present in grains and alike to that present in meat (Singh and Basu, 2012).

The common pulses used as food are chickpea (*Cicer arientinum* L.), pigeon pea (*Cajanus cajan*), black gram (*Phasedous mungo*), green gram (*Vigna radiata*), lentil (*Lens culinaris medic*), cow pea (*Vigna unguiculata*), kidney beans (*Phaseolus vulgaris*), winged beans (*Psophocarpus tetragonolobus*), horse gram (*Doliches biflorus*), moth bean (*Vigna aconitifolia*), fava beans (*Vicia faba* L.), grain amaranth (*Amaranthus* spp.), jack bean (*Canavalia gliadata*) and grass peas (*Lathyrus sativus*). Besides protein, pulses are also a rich source of energy, phosphorous and calcium (Table 1). Studies showed that the consumption of pulses may reduce the risk of cardiovascular disease (Flight and Clifton, 2006), and diabetes, and may have protective effects against numerous types of cancers such as breast cancer, colon and rectal cancer (Campos-Vega et al., 2010).

Pulses contain complex carbohydrates (oligosaccharides, dietary fibers and resistant starch), proteins with a good amino acid profile (high in lysine content). They also contain important vitamins, such as water-soluble B vitamins (thiamine, riboflavin, niacin, and pyridoxine), folates, ascorbic acid and tocopherols, minerals as well as antioxidants, polyphenols and other phytochemicals (Tiwari et al., 2011). The pulses are low in sodium but are the good source of minerals, such as calcium, copper, iron, magnesium, phosphorus, potassium and zinc (Venter and Eyssen, 2001). The predominant fatty acid in the pulses is linoleic acid, although they also contain α-linolenic acid (Venter and Eyssen, 2001).

Table 1: Nutritional composition of major pulses.

S. No.	Common name	Latin name	Energy (Kcals)	Moisture (%)	Protein (%)	Fat (%)	Carbohydrate (%)	Mineral (%)	Fibre (%)	Calcium (mg/100g)	Phosphorus (mg/100g)	Iron (mg/100g)
1	Pigeon pea (tur, red gram, arhar)	*Cajanus cajan*	301	10	21	1.25	58	4.1	10.4	143	228	3.6
2	Green gram (moong)	*Vigna radiata*	334	10	24	1	57	3	4	124	326	4
3	Black gram (urad, urd)	*Vigna mungo*	-	-	23.6	0.45	67		3.5	-	-	-
4	Chick pea (bengal gram, chana)	*Cicer arietinum*	360	10	18.5	4.3	55	3.5	4	138	364	5
5	Kabuli Chick pea (kabuli chana)	*Cicer arietinum*	-	10	20.5	5.7	55	3.8	4	68	337	3.8
6	Lentil (masoor)	*Lens culinaris*	343	12	25	1	59	2	1	69	293	7
7	Field pea (dry pea)	*Pisum sativum*	315	16	20	1	56	2	4	75	298	7
8	Kidney bean	*Phaseolus vulgaris*	333	11.8	23.6	0.8	60	2.1	24.9	143	407	8.2
9	Cowpea (black eyed pea, southern pea)	*Vigna sinensis*	323	13	24	1	54	3	3	77	414	-
10	Faba bean (bell bean, horse bean, broad bean)	*Vicia faba*	341	11	26.12	1.53	58.29	-	2.5	103	421	6.7

(Source: Vishwakarma et al., 2017)

Amino acid composition of pulse protein

Pulses contain almost all the essential amino acids. Unlike cereals which is deficient in lysine, pulses are deficient in methionine, tryptophan and cysteine as it can be seen from Table 2.

Table 2: Amino acid profile of major pulse proteins

Amino Acid	Groundnut	Soybean	Pea	Chickpea	Lentil	Grass pea
Arginine	11.0	7.4	6.84	0.48	9.10	2.18
Cystine	0.9	1.6	1.55	ND	0.44	0.32
Glycine	6.0	4.5	4.32	0.26	10.22	0.99
Histidine	2.5	2.4	2.52	0.25	6.84	0.63
Isoleucine	3.0	4.6	3.33	0.36	5.06	2.69
Leucine	6.1	7.8	6.58	0.48	8.09	
Lysine	3.6	6.1	6.84	0.91	5.69	1.57
Methionine	0.4	1.4	1.03	0.12	1.18	0.30
Phenyl alanine	4.9	5.5	4.19	0.42	5.55	1.27
Threonine	2.8	3.8	3.59	0.06	5.62	0.80
Tryptophan	ND	1.3	0.94	ND	ND	ND
Tyrosine	3.7	3.5	3.16	0.19	5.05	0.59
Valine	3.7	5.2	3.89	0.38	7.24	0.99

Source: Srilakshmi 2019

Anti-nutritional compounds

Pulses contain a number of basic nutrients and some other chemical compounds produced and accumulated during the growth of the seed to defend against herbivorous animals, microorganisms and viruses (Soetan and Oyewole, 2009). Some of these chemicals are biologically active and called as secondary metabolites. The secondary metabolites affect the nutritional quality of pulses and referred as antinutritional factors. These compounds are sometimes toxic, unpalatable or antinutritive due to their abilities to block nutrients, inhibit metabolism or reduce digestion (Enneking, & Wink, 2000). These anti-nutritional components are limiting factor for the food use of pulses. The deficiency of pulses in sulfur-containing amino acids and low protein digestibility are the other factors which limit their food use.

The anti-nutritional factors include flatulence causing raffinose family oligosaccharides (α-galactosides), phytates, phenolics, tannins, cyanogens, enzyme (trypsin, chymotrypsin and α-amylase) inhibitors and lectins (phytohemagglutinins). Only a few pulses contain all these antinutritive factors, most of the pulses contain some of them. The composition of anti-nutritional compounds present in different pulses is given in Table 3. The high polyphenolic containing pulses are the dark, highly pigmented varieties, such as pigeon pea, kidney beans, black gram and chickpea. Condensed tannins (proanthocyanidins) are mainly concentrated in hulls of pulses.

Trypsin and chymotrypsin inhibitors present in pulses are active against proteases, amylases, lipases, glycosidases, and phosphatases. Pulses, especially beans, are the second largest group of seeds after cereals as natural sources of a-amylase inhibitors (Campos-Vega et al., 2010). Though haemagglutinins or lectin are present in most foods, the pulses are the main sources of lectins in human food. Beans are an important source of lectins, but some cultivars may have higher lectin content than others. Lectin is one of the major proteins found in all pulses except chickpea (Table 3). All lectins bind to one transition ion, usually manganese, and one calcium ion. Phytate or phytic acid is present in all pulses and it is located in the protein bodies in the endosperm. It is present in the form of Inositol phosphate (Ins) in pulses. The most abundant inositol phosphate in dry pulses is InsP6, accounting for about 83% of the total inositol phosphates. The InsP6 concentration is higher in dry beans and pigeon peas than in lentils, field peas, and chickpea. Oomah et al. (2008) reported that phytic acid, expressed as InsP6, represents 75% of the total phosphorus in several Canadian bean varieties. Tannins content of pulses vary from 0.0 to 2.0% depending on the species and seed coat colour. It is believed that high tannin pulse varieties are of lower nutritional quality than low tannin varieties (Reddy et al., 1985). Red kidney bean has highest tannin content followed by the black gram, pigeon pea and lentils (Table 3).

Table 3: Major antinutritional compounds content in different pulses (Dry matter basis)

Pulse	Anti-nutritional compounds									In-vitro protein digestibility (%)
	Enzymatic inhibitors			Non-enzymatic inhibitors						
	Trypsin inhibitor (TIU/mg protein)	Chymo-trypsin inhibitor (IU/mg)	α-amylase inhibitor (IU/g)	Heamag-glutinin activity (HU/mg)	Tannins (mg/g)	Phytic acid (mg/g)	Total oxalate (mg/100g)	Total oligo-saccharide (g/100g)	Total saponin (mg SBaE/g)	
Green gram[1]	15.8	-	-	26.70	3.30	5.80-12[a]	-	-	2.20	67.2-72.2[a]
Chick pea[2]	8.10-20.89	6.1-8.8	3.1-10.5	6.22[b]	0.04-4.85	5.8-12.1	70.3[b]	2.94[b]	0.91-60	65.3-72.2[a]
Pigeon pea[3]	8.1-31.28	2.1-3.6	22.5-34.2	25.01	0-6.88	5.01-12.7[a]	-	1.84-3.35	21.6-34.9	60.4-74.4[a]
Lentil[4]	3.6-7.6	0[b]	0[b]	50[b]	1.28-3.9	4.11-12.5	13.3-144	4.75[b]	34[b]	
Black gram[5a]	7.1-10.6	-	-	-	5.3-9.2	3.7-13.7[a]	-	-	8.7-21.4	55.7-63.3[a]
Field pea[6]	0.78-6.32	2.73-4.85	16.8[b]	80[b]	2.78-3.09	3.0-13.3	0-103	5.63-5.73	0.01-31.7	83.7
Kidney bean[7]	3.10-31.3	3.97[b]	76-675	74.5[b]	0.03-19.9	9.9-29.3	13.9-946	6-6.19	0.02-16	68.1
Cow pea[8]	3.4-67.1	1.6	1.4-89.5	40-640	-	2.64-15.24	42-770	-	2.51-37.16	-
Faba bean[9]	2.31-7.20	3.56[b]	18.9[b]	49.3[b]	0.13-21	6.4-21.7	194[b]	-	3.52[b]	70.8

[1]Mubarak (2005); [2]El-Adawy (2002); [3]Singh (1988); Solomon et al. (2017); [4]Hefnawy (2011); [5]Suneja et al. (2011); [6]Khattab and Arntfield (2009); [7]Yasmin et al. (2008); [8]Gonçalves et al. (2016); [9]Alonso et al.(2000); [a]Chitra et al. (1995); [b]Shi (2015)

Processing and utilization

Pulses have been assimilated in different forms of traditional and staple diets to supplement basic protein and energy requirements and provide functional properties, beneficial to human health (Rehman et al., 2001; Frias et al., 2016; Benouis, 2017). There are many processing methods that enhance the nutritional content and bioavailability of many vitamins and minerals in the pulses, such as fermentation, germination, parching etc. Also, in pulses, some anti-nutritional compounds are present, which on consumption can cause various digestive problems, reduces bioavailability of nutrients and can cause serious health problems in many cases. Many of these anti-nutritional compounds get removed during cooking and other processing techniques (discussed later in this chapter). The various processing techniques includes boiling, grinding, parching, toasting, roasting, puffing, germination and fermentation. Pretreatment steps such as boiling, soaking, or roasting is employed to facilitate the husk removal that can be accomplished by subsequent pounding, grinding, or milling. The various processing techniques, their function and effects on quality of pulses are summarized in Table 4. Different processed and food form of legumes are presented in Table 5.

Table 4: Processing techniques, their functions and effect.

Processing technique	Function	Effect
Decortication/milling	Removal of outer husk from whole pulse seeds by pounding and grinding	Removal of husk, easier digestibility, shorter cooking time.
Soaking	Dipping in water for >6 hours	Reduces anti-nutrient content, shorter cooking time, facilitate removal of husk, improves milling.
Germination	Sprouting the seeds by soaking the whole unhusked grains for 24 hours, and then spreading on a damp cloth for up to 48 hours	Improves vitamins content, protein digestibility is improved, anti-nutritional factors are reduced.
Fermentation	Soaking for prolonged periods to allow to ferment	Carbohydrates get converted to simple forms, nutritive value increases, digestive value increases.
Roasting & Parching	Heat treatment without oil/fat	Improves palatability, crispiness, easier decortication.

Table 5: Different processed and food form of legumes

Processing method	Processed form	Food form
Pounding, grinding, milling	Meal, flour, grits, paste, dhal	Unleavened breads, biscuits, cakes, noodles, porridges, gruels, stews, sauces etc
Boiling	Whole or de-husked	Stews, vegetable dishes, soups, condiments

Processing method	Processed form	Food form
Roasting, parching	Whole or de-husked	Snacks, garnish
Frying	Whole, de-husked, flour, paste, batter	Snacks, used in bakery items
Puffing	Whole or de-husked	Snacks
Germination	Whole or de-husked	Bean sprouts, curry, vegetable dishes
Fermentation	Flour, paste, batter	Fried doughs, oriental foods, condiments
Agglomeration	Flour	Cereals, cous-cous salad
Steamed	Whole or de-husked	Vegetable dishes, salads
Mechanical decortication and milling	Whole legumes, flour, grits, paste	Traditional foods: porridge, stew, gruel, cous-cous, etc. New foods: ethnic breads, pasta products, snack foods (e.g. chips), confectionaries
Precooking, drying, grinding	Whole or dehusked legumes, quick cooking legumes, instant legume powders	Traditional foods as above beverages, snack foods and dips, soups, oriental foods
Agglomeration, air classification, precipitation	Legume protein concentrates	Legume-cereal mixtures: ethnic breads, baked goods, pasta products, infant foods, cakes, pancakes, simulated foods: rice/meat analogs, meat extenders, snack foods

Milling

In broad terms 'milling' is defined as a process by which materials are reduced from a larger size to a smaller size (Wood and Malcolmson, 2011). However, in case of pulses, milling is usually referred as removal of seed coat. Removal of seed coat to produce polished seed (dehulling) and cleavage of the two cotyledons to produce split seeds (splitting, also known as *dhal* in India) are the most common milling operations in Asia and Africa (Wood and Malcolmson, 2011). About 80-90% potential anti-nutritional polyphenols are reported to be present in the seed coat, which are significantly reduced by dehulling operation (Singh, 1993). The main objective of pulse milling is to completely remove the hull from pulse seeds while minimizing the production of powder, broken, and in certain pulses split seeds (Wang et al., 2005). Higher ratio of whole dehulled seeds to split seeds is desired in case of pigeon pea and lentil as whole dehulled seeds are the most valuable fraction for processors (Vandenberg, 2009). The sequence of unit operations involved in pulse milling is cleaning and grading, pitting, pre-milling treatment, drying, dehulling, and splitting and discussed below.

Pitting

Pitting or surface scratching is one of the basic unit operations prior to application of any pre-milling treatments in pulse milling (Narasimha et al., 2002). It is reported that passing the seeds quickly through the roller mill results in scratches or cracks in hull and removal of waxy layer of seed coat (Sokhansanj and Patil, 2003) and improves the absorption of pre-treatment agents (Narasimha et al., 2002). Dedicated machine for pitting of pulses is not reported in literature and pulse dehulling machines are generally used for pitting.

Pre-milling treatments

The pre-milling treatments are generally employed to loosen the seed coat (Singh, 1995). It is aimed to reduce breakage and improve quality of end product. Various methods of pre-milling treatments are developed for commercial and small-scale milling of pulses such as pre-treatment with water, edible oils, chemicals, enzymes and a combination of heat and water (hydrothermal).

Moisture

It is oldest pre-milling treatment method for loosening the bond between hull and cotyledons. The process involves soaking seeds in water for 6-8 h; mixing with red earth slurry (3%) in some regions of India; sun drying for 2-4 days; tempering for 8-12 h; separation of red earth and milling (Lal and Verma, 2007). Soaking in water results in expansion of seeds and when the seed is dried, the cotyledons shrink more than the hull resulting in a bubbled hull, which can be removed during dehulling (Sokhansanj and Patil, 2003). Moreover, when grains are soaked for 6-12 h and dried, the cotyledons cave in at their surface of fusion and touch each other at the periphery. The wet method has the advantage of facilitating good dehulling and splitting of the cotyledons with less breakage though it adversely affects the cooking quality, labour intensive and dependent on climate. Small scale pulse dehulling units practice alternate wetting and sun drying of pigeon pea followed by oil application and dehulling (Goyal et al., 2005).

Edible oils

At present edible oil pre-milling treatments are commonly practiced in commercial milling of pulses. The process involves pitting of seeds (at 10-12% moisture content) application of any common edible oil @ 0.1-1% (w/w), sun drying in thin layer for 2-5 days, mixing small amount of water (to increase moisture content by 2-5%), tempering in heaped form for 12-18 h, sundried to about 10-11% moisture content, and milling (Sokhansanj and Patil, 2003; Tiwari et al., 2007). The oil is believed to penetrate through the hull that helps in release of its binding with cotyledons. This method is said to produce dhal that cooks faster than the dhal produced by the wet method. The major disadvantage of this method is high dehulling losses due to breakage and powdering.

Chemicals

Aqueous solution of various chemicals such as acetic acid, NaOH, $NaHCO_3$, Na_2CO_3, NaCl etc were reported for loosening the bond between hull and cotyledons of pigeon pea. Considerable improvement in dhal yield can be obtained after treatment with $NaHCO_3$ solution. Hydrolysis of pectin and gum present in seeds was reported to be the possible reason for increase in dehulling efficiency (Srivastava et al., 1988). However, leaching of sugars, proteins, and electrolytes into the soaking medium (loss of nutrition), discoloration of *dhal*, and non-availability of by-products (hull, broken and powder) for cattle feed due to residual chemical are the main detrimental aspects of this pretreatment (Lal and Verma, 2007) and therefore this process was not adopted for commercial processing.

Enzymes

The use of enzyme for partial hydrolysis of cell walls was initially explored for improvement of oil extraction from canola (Sosulski et al. 1988). This approach was then evaluated for partial hydrolysis of hull by enzymatic reactions to facilitate dehulling of pigeon pea (Verma et al., 1993). The pretreatment with crude enzyme produced by *Aspergillus fumigates* resulted in 86.74% dehulling efficiency for pigeon pea (Verma et al., 1993). To identify the specific enzymes facilitating dehulling of pulses, Sreerama et al. (2009) evaluated xylanase and protease for dehulling properties of green gram, black gram, pigeon pea and horse gram. They reported that xylanase mediated degradation of non-sugars poly-saccharide facilitated easy dehulling of pulses. Protease pretreatment is more efficient in improving the dehulling properties of pigeon pea, green gram and black gram. The partial hydrolysis of these proteins by protease facilitate the easy removal of hull. However, the application of enzymes to treat the seeds for easy dehulling of pulses is still at laboratory stage.

Hydrothermal

Heating of pigeon pea for easy dehulling was practised earlier in some parts of India. The heat treatment based pre-milling treatment process was then standardized by Central Food Technological Research Institute (CFTRI) India for pigeon pea milling. The process comprised pitting, oil application and conditioning for 6-8 h, heating grains at 120-180°C in LSU dryer for some time and conditioning for 5-6 h in the drying bin, heating grains again for some time and conditioning for 10-12 h in the drying bin and dehulling with abrasive mill (Singh and Ilyas, 1987). Dhal yield of 81-86% have been reported for different pulses. It is believed that the grain sweats during the conditioning process and hull is loosened (Goyal et al., 2005). This practise is now followed for dehulling of peas at commercial scale. Heating with microwave was investigated to improve the dehulling characteristics of the black gram (Joyner and Yadav, 2015). Microwave dosage of about 972 J/g was reported to be optimum for dehulling at a rate of 630 W or higher power level with dehulling efficiency of 73.7%

and little change in colour of dehulled material. However, hydrothermal treatment of pulses to improve dehulling is still at research stage and needs further investigations.

Dehulling machines

Dehulling of pulses is usually done with the application of abrasive forces (Sokhansanj and Patil, 2003). In early seventies, vertical stone mills were commonly used for dehulling pulses in India, but due to high grain damage (up to 20-45%) and poor product quality, stone mills are not in use. Several machines have been developed for dehulling of pulses. Most common commercial machine for pulse dehulling is emery coated cylinder-concave system. Emery grits of 14-16 to 36-40 mesh sizes (BSS) are used in dehulling machines (Narasimha et al., 2003). Fine-mesh emery is used for pitting whereas large grit emery (14-16 mesh) is used for dehulling, which may vary according to type of pulses. Pulse dehulling machines developed by different organizations are the modified version of commercial dehulling machines and work on the principle of abrasion. A comparative performance of different dehulling machines is reported in Table 6. Invariably the dehulling efficiencies of three dehulling machines namely IIPR dhal chakki, CIAE Dhal mill and CFTRI dhal mill were 49.53%, 40.35% and 45.35%, respectively for untreated pigeon pea grains (Lal and Verma, 2007). Hiregoudar et al. (2014) also reported dehulling efficiency of 43.3% for untreated pigeon pea when dehulled with CFTRI mini dhal mill. It may therefore be inferred that pretreatment of pulses influence the performance of dehulling machine rather than the type of system used for application of abrasive force.

Table 6: Dehulling machines for pulses, their mechanism and performance.

S. No.	Dehulling machine	Mechanism	Functional unit	Capacity (kg/h)	Pretreatment method	Dehulling efficiency (max., %)	Dhal yield (%)	Reference
1.	CIAE dhal mill	Abrasion	Emery roller concave	100	Wet or oil	80	77	Lal and Verma, 2007
2.	PKV mini dhal mill	Abrasion	Emery roller concave	125-150	Oil	83	74	Goyal et al., 2005
3.	CFTRI mini dhal mill	Abrasion and shear	Truncated cone type emery concave	100-150	Oil and heating	87	80	Lal and Verma, 2007
4.	IIPR dhal chakki	Abrasion and impact	Vertical plate mill (metal and rubber)	75-80	Oil	88	76	Lal and Verma, 2007
5.	Pantnagar dhal mill	Abrasion	Emery roller concave	75-80	$NaHCO_3$ solution	80	80	Goyal et al., 2005

S. No.	Dehulling machine	Mechanism	Functional unit	Capacity (kg/h)	Pretreatment method	Dehulling efficiency (max., %)	Dhal yield (%)	Reference
6.	IARI mini dhal mill	Abrasion	Emery roller concave	120-140	Wet	80	83	Goyal et al., 2005
7.	PRL batch dehuller	Abrasion by abrasive discs	Emery disc concave	7	-	-	-	Sokhansanj and Patil, 2003
8.	Schule dehulling machine	Abrasion by abrasive discs	Vertically placed emery discs discs cylinder	16000	-	-	-	Scheule, 2016
9.	Decomatic Bernhard Keller KG vertical roller mill	Abrasion by abrasive discs	Rotating stones inside wire mesh cage	-	-	84.4	-	Sokhansanj and Patil, 2003
10.	S. K. Engineering and Allied Works, India	Abrasion	Truncated cone type emery concave	500-2000	Oil and heating	85	-	Sokhansanj and Patil, 2003
11.	ICRISAT mini dehulling mill	Abrasion and impact	Horizontal stone plates	40	Oil	71	-	Goyal et al., 2005
12.	TNAU dhal mill	Abrasion and impact	Horizontal stone plates	-	Wet	-	-	Goyal et al., 2005

Source: Vishwakarma et al 2017

Milling losses

Narasimha et al., 2003 reported that average hull content of pulses ranges from 4.9% for *kabuli* chick pea to 14.4% for pigeon pea. Therefore, the feasible theoretical *dhal* yield, which includes dehulled whole cotyledons and dehulled splits, is 84.6% to 95.1%. In small capacity mills, 74-83% *dhal* yield has been reported for graded and pretreated pigeon pea (Narasimha et al., 2003). Goyal et al. (2005) reported 68-76% *dhal* yield in commercial milling of pigeon pea and remaining 8.6-16.6% cotyledons are converted into broken and powder. Formation of broken and powder is considered as milling loss because these fractions are not used for human consumption. Singh et al. (1992) reported be 24.6% and 9% broken formations during dehulling of wet treated pigeon pea with horizontal burr mill and abrasive machines, respectively. Ramakrishnaiah et al. (1993) reported 8-20% broken and 8-20% powder having hull also during milling of different pulses using small scale abrasive mills in India.

Pulse milling by-products

The major portion of the pulses processed is milled by the *dal* mills with daily capacity ranging from 0.5 ton to 10 tons per day. The pulses contain 11-14% husk and 2-5% germ and rest are the seed endosperm. The extraction rates of processing are between 70 and 88% of raw material. The main by-products of pulse milling are in the form of broken (6- 13%), mixture of germ and powder (7-12%) and husk (4-14%). Small broken and husk are used as cattle feed. Husks of lentils are used in poultry feed, broken of Bengal gram are fed to horses and used in *besan* preparations. Broken of pulses are milled to produce flour and are used in *papad* preparations also.

Soaking

Soaking works as a pretreatment to decortication by softening the husk for its easy removal. Legumes/pulses may be steeped/soaked into hot water below its boiling point to allow the seed coat to swell and loosen from the cotyledon. Soaking reduces the water soluble vitamins (loss of vitamin B_1 or thiamine is more in alkaline medium) but significantly improves the protein digestibility, starch digestibility, availability of minerals like zinc, iron and dietary fiber components such as cellulose, hemicellulose, lignin and pectin which vary in different genotypes of pulses. Soaking during germination facilitate the synthesis of Vitamin E. Furthermore, soaking tends to minimize anti-nutritional factors like trypsin inhibitor, phytates, tannins, saponins, β-ODAP (β-N-oxalyl 2,3 di-amino propionic acid) and total phenol etc. (Vasishtha and Srivastava, 2013; Srivastava and Srivastava, 2006). However, soaking triggers synthesis of phytase enzyme, which results in leaching out of phytic acid; break down of oligosaccharides attached to aglycone, which support softening of tissues; polyphenols oxidase enzyme, which are activated, and results in degradation and loss of polyphenols. The most common and simplest home-scale method for decortication of pulses and legumes is by pounding or grinding, or a combination of these two methods. This involves pounding of grains for dehusking by using a mortar and pestle after mixing with small quantity of water and drying in the sun for a few hours. This system of hull removal can be classified into two categories, namely dry method and wet method. A combined wet and dry method is also used. Both methods are practiced throughout the semi-arid tropical zone, the former having greater acceptance in Central and Northern India, Africa, and Central America.

Boiling

Boiling is a method of cooking in which pulses are kept in water at temperature higher than 90 °C for a certain period which helps to cook and soften the seeds, making it edible for consumption and also facilitates removal or inactivation of certain ant-nutritional constituents. In India, *dhals* are often cooked until soft, mashed, then mixed with water and re-boiled to give a consistency of a soup or gruel. Uncooked legume seeds contain anti-nutritional factors that can be toxic if large amounts are consumed. Trypsin inhibitors and haemagglutinins disappear at 90 °C, however, polyphenolic compounds content decreases. Quality of protein is improved more by moist heat than by dry heat treatment in pulses, as available lysine is decreased in roasted pulses as compared to boiled and pressure-cooked ones. Heat treatment causes loss of methionine, the most important amino acid of legumes. Cooking has slight effect on calcium, magnesium and total iron content of pulses. Loss of thiamine may occur due to heat applied. Steam helps greater heat penetration and save cooking time. Now a days *dhals* are cooked mostly under pressure.

Germination

Germination (sprouting) is a traditional, non-thermal process which improves the nutritional quality of cereals and pulses by increasing nutrient digestibility, reducing the levels or activities of anti-nutritional compounds, boosting the contents of free amino acids and available carbohydrate, and improving functionality (Ridout et al., 2002; Vidal-Valverde et al., 2002; Frias et al., 2002). Amongst the functional properties, solubility is often considered the most critical because it affects other properties such as emulsification, foaming and gelation. It plays an important role in the reduction of anti-nutritional factors, including phenolic compounds, phytic acid, trypsin inhibitors and oligosaccharides (Shimelis and Rakshit, 2007). Most of the anti-nutrients bind to protein and/or form complexes with enzymes, rendering them unavailable or inactive for digestion. Germination involves chemical changes such as the hydrolysis of starch, protein and fat by amylolytic, proteolytic and lipolytic enzymes, respectively. When grains are hydrated (soaked) and held (sprouted) under ambient conditions, both endogenous and newly synthesized enzymes begin to modify seed constituents. Thus, complex macromolecules are broken down into lower molecular weight molecules which are more digestible and readily absorbed by the body (Dhaliwal and Aggarwal, 1999). Luo et al (2009) reported that with increased germination time at 30 °C, phytic acid progressively decreased from 9 to 69% in faba bean.

Fermentation

Fermentation is perhaps one of the oldest methods for processing food grain legumes. In practice, the fermentation process breaks down carbohydrate (starch) to acid as the final end product by the action of microorganisms (bacteria, molds, and yeast). In the household practice, microorganisms present in the atmosphere are the fermenting organisms. This is also true in village-scale operations, conversely, controlled fermentation, using specific molds and bacteria, is followed in large-scale commercial operations. The main effect of fermentation, regardless of the fermenting organism used, is to make more of the grain nutrients available for assimilation in the body. In this respect, the digestibility of the legume protein is increased. Luo et al (2009) observed that fermentation treatments can decrease phytic acid (48–84%) from faba bean, followed by soaking at 10 °C after preheating (36–51%). Using this method, more protein in the form of amino acids is readily absorbed and utilized. The natural formation of antioxidants during the fermentation process has practical significance in tropical regions. In addition, fermented products have an increased storage life at room temperature, since organic acids and the amino acids produced during the process can inhibit the growth of pathogenic microorganisms. The major fermented products prepared from pulses are presented in Table 7.

Roasting

Roasting, sometime referred to as toasting or parching, is practiced mainly in India and Africa. Roasting refers to the method in which usually whole, or de-husked grains are exposed to dry heat. This is accomplished either directly by placing the whole grains and beans at the edge of a fire, or directly upon it, or in hot ashes or sand in contact with the fire. Roasted leguminous seeds with husk are commonly consumed as snacks. The roasting process improves the flavour, texture, and nutritive value. Puffed grain legumes are prepared at household level in the similar way that is used for roasting. Puffing brings about a light and porous texture in split dehusked *dhal.* For puffing, grains are soaked in water and mixed with hot sand, which has been heated to about 250°C and then toasted for a short period, approximately 15-25 seconds. After the sand is sieved off, the grains are dehusked between a hot plate and rough roller. The more common legumes prepared in this manner are peas and chickpeas. Frying is mostly used on previously processed legumes, which are in the form of a flour, paste, batter, or dough. In India, chickpea, black gram *(Phaseolus mungo),* and peas *(Pisum sativum)* are often prepared into doughs or paste, which are deep fried into crispy products. In addition, mung bean or its splits is also fried in a little fat and eaten as a snack.

Table 7: Major fermented products prepared from pulses

Product name	Produce	Country of origin	Microorganism involved	Mode of consumption/form	References
Idli	Black gram	India, Sri Lanka	*L. delbrueckii, L. fermentum, Lactococcus lactis, Leuc. mesenteroides, Strep. lactic Ped. cerevisiae*	Breakfast food	Nagaraju and Manohar, 2000
Dhokla	Black gram	India	*B. cereus, Ent. faecalis, Leuc. mesenteroides, L. fermenti, Tor. candida, Tor. pullulans*	Breakfast food, Snack	Sands and Hankin, 1974
Dosa	Black gram	India	*Bacillus* sp., *L. fermentum, Leuc. mesenteroides, Streptococcus faecalis*	Breakfast or snack food	Iwuoha and Eke, 1996
Dawa Dawa	Local pulses	West and Central Africa	*B. licheniformis, B. subtilis*	Condiment, meat substitute	Pierson et al., 1986
Papad	Bengal gram, black gram, lentil, red or green gram	India	*C. krusei, S. cerevisiae.*	Condiment or savory food	Vidal-Valverde et al., 2002
Wadies	Black gram and oil	India	*L. fermentum, L. mesenteroides*	Spicy condiment or an adjunct for cooking vegetables or rice	Köksel et al., 1998
Tempeh	Soybean	Indonesia, New Guinea, Surinam	*Asp. oryzae, Rhiz. oligosporus*	Breakfast food or snack	Patwardhan, 1962

Convenience Foods

Today, at any given time, tens of thousands of various types of convenience foods are offered for sale to consumers throughout the world. All of them offer the advantages of food to consumers that is readily available, convenient, inexpensive, and often of surprisingly pleasing organoleptic quality, requiring only minimal or no preparation before consumption. Convenience foods can be looked at in several ways. These foods can be narrowly classified by the processing technologies employed: canning, freezing, dehydration, chilling, chemical preserving, etc., or by the type of food: frozen and canned vegetables; cake mixes and bakery products; soups, sauces, and condiments; processed meats and fish; chilled and frozen dairy-based products; ready-to-eat and shelf-stable dishes; plus many other types. The first commonly available convenience foods were canned goods, which were developed in the 19th century for military purposes. The little tins made it easy to store, transport and prepare food on the battleground. Various development has also been happened in convenience foods from pulses like quick cooking pulse, canned pulses, pulses in puffed snacks, pulse flakes etc.

Quick cooking pulses

The technology for quick cooking pulses has significant application for its consumption since their preparation requires shorter cooking time. The resulting products, however, possess similar or improved physical, chemical, and nutritional properties as compared to those prepared by standard, long-time cooking processes. Its development came about during war time when a shortage of tin cans existed. Cooking of seeds by traditional method requires prolonged time, which involves cost in terms of fuel, energy and also affects the nutritional quality adversely. Quick cooking pulses are especially suited for operational pack rations of armed forces because of their light weight, easy cooking characteristics and long shelf life. Greater emphasis is being given for marketing the new products of legumes such as quick cooking dhal or instant dhal of grain legumes which have good market potential (Singh 2007). Boiling or simmering, if necessary, should not exceed more than 5 min to bring any dehydrated food product into ready to serve form. Reduction in cooking time of pulses by a simple process is requirement of today's consumer.

Several special techniques have been tried by the researchers to produce various quick-cooking dehydrated foods. In the case of pulses and beans subjecting the cooked grains to the action of proteolytic enzyme, papain was used by Bhatia, et al (1967) to produce quick cooking dehydrated products. Attempts have been made to minimize shrinkage and improve rehydration characteristics by pretreatments with additives like glycerol prior to drying and flashing techniques such as explosive puffing, vacuum puffing and centrifugal fluidized bed drying. The latter is capital intensive. The technique of high temperature short time (HTST), pneumatic drying have been applied to a variety of pulses (Jayaraman et al 1980 & 1981) to bring about porosity in the products thereby reducing their drying and rehydration time considerably with significant improvements in their texture and rehydration characteristics. This was achieved by exposing the cooked pulses (whole grams and ***dals)*** and blanched vegetable pieces (dice or strips) initially to air at a high temperature (160-200°C) for a short period (4-8 min) in a HTST pneumatic drier followed by conventional method of drying at 60-70°C in a tray or fluidized bed drier (Jayaraman K 1984). Considerable reduction in drying and rehydration times could be achieved by the HTST process. The HTST treatment does not adversely affect the shelf stability of the pulses and enhances porosity. The technique involves simple equipment and less capital investment as compared to other methods of drying, such as explosive puffing, achieving the same objective. It is amenable to continuous processing as the time of high temperature drying is very short (Jayaraman, 1980).

Bede, (2007) determined the cooking time of three varieties of cowpea (*Vigna unguiculata*) by sudden and rapid decrease in the temperature (quenching) of boiling cowpea seeds, by the addition of water at a lower temperature. The results showed that quenching with water at a temperature of about 62 °C produced the maximum reduction in the cooking time for all the varieties. A reduction in cooking time of cowpea in the range of 13-29% and 25-40% in case of pigeon pea (*Cajanus cajan*) can be obtained by following above methods. Shruti et al (2011) prepared quick cooking dhal from pigeon pea (variety UPAS 120) following milling, pre-treatment with sodium chloride solution (1%), flaking and drying. The authors concluded that treatment with sodium chloride followed by flaking is effective in reducing the cooking time of pigeon pea dhal. Nayak and Samuel (2015) prepared instant pigeon pea dhal by soaking in sodium chloride solution, cooking in pressure cooker, partial drying, flaking and final drying. The optimum level of salt concentration (%), cooking time (minute) and flaking thickness (mm) was reported as 0.75, 12 and 0.75 respectively. The instant pigeon pea dal thus prepared could be reconstituted in less than 6.16 minutes. Fiave, (2018) prepared quick cooking cowpea through soaking, steam cooking followed by oven drying. He observed significant reduction in the cooking time of the dried precooked cowpeas. Pre-cooked and dried pulses have been reported to undergo rapid auto-oxidation during storage leading to disappearance of yellow colour and formation of off flavour (Patki and Arya 1994).

Canned beans

Canning represents the most common method of processing legumes for human consumption, especially in the developed countries. It is done on a commercial basis as the process itself involves considerable time and the need for cooking kettles (pressure cookers) and can-sealing equipment. Commercial legume canning operations are common in North America and, to a lesser level in Central and South America, Africa, and Asia. In most cases, canned legume products in developing countries are consumed by the higher-income class, or are exported to developed countries, since they are higher in cost than traditionally processed legumes. The most popular kinds of legumes used for canning are beans and belong to the *Phaseolus* genus, namely, navy or kidney beans *(P. vulgaris)* and lima or butter beans *(P. lunatus).* The former bean primarily known as the common bean, is used in the preparation of canned North American-style "baked beans." Canned kidney beans usually consumed as a vegetable side dish or they may be used as the basic ingredient in a salad.

A new canning operation was established for pigeon peas which initially uses dried, whole beans, which are washed in cold water. After draining, the beans are allowed to soak overnight in water, during which the moisture content of the dried beans increases from an initial 10-12% to approximately 20%. This facilitates cooking of the whole bean, as there is a softening of the beans. This step is essential because the beans are processed as whole. Afterwards, weighed amount of beans are put in each can, a liquid is added to the can, this being either a thin sauce (tomato) in the case of navy beans, or water, in the case of peas. The cans are then sealed and subsequently placed in a retort for cooking. The heat-processing step proceeds for a time that is dependent upon the temperature and pressure used. Since the retort is essentially a large pressure cooker, a shorter cooking time for processing the beans can be used than cooking at atmospheric pressure. Usually, 90-minute process is used at a temperature of 121°C (250 °F). After retorting, the cans are cooled in the retort under pressure for approximately 15 minutes prior to their removal. This prevents overcooking of the canned beans, even though the steam processing step has been completed. In addition, the cooling step prevents the cans from bursting when they are exposed to atmospheric pressure. Research on canning beans has concentrated on studying processing time as it relates to the final texture and consistency of the products and nutritional losses involved in the canning process. It is widely recognized that substantial losses of some water-soluble vitamins can occur during the presoaking (hydration) step, or this loss can occur if the liquid is discarded in which the beans have been cooked. According to reports as much as 50% of the thiamine can be lost during canning (Berrios, 2006). Significant amounts of riboflavin, niacin, and vitamin A have been found in the liquid medium surrounding the beans during the hydration and cooking steps. The beneficial effects of heat-processing on eliminating some of the anti-nutritional factors present in beans is overshadowed by the concurrent losses of protein quality in the canned beans. The lower availability of lysine is partly responsible for this effect. Furthermore, soaked samples have shown a reduction in the nutritive value with a cooking time higher than 10 minutes. The determination of the optimum cooking time and temperature for canning of beans to minimize protein quality loss is a continuing research effort.

Expanded food fortified with pulses

Extrusion cooking technology is a high-temperature short-time (HTST), versatile and modern food operation that converts agricultural commodities, usually in a granular or powdered form, into fully cooked food products. Extrusion is a mechanical process used to create objects or products of desired shapes, sizes, and texture by forcing material through an orifice ("die" opening) under pressure. In general, the final extrudate has low moisture and is considered to be a shelf-stable food product (Berrios et al., 2010). The HTST cooking process significantly reduces microbial population, inactivates enzymes, and minimizes nutrient and flavor losses in the food being produced (Yadav et al 2013). HTST extrusion cooking is used by many sectors of the food industry to produce expanded snack foods (Yadav et al 2013), ready-to-eat (RTE) (Sharma et al 2015), breakfast cereals (Williams et al., 2010) etc. Crispness, a typical quality attribute of all these foods, is directly related to expansion, which in turn is dictated by various extrusion parameters (Alavi, 2004).

Conventional food extrudates contain significant concentrations of starch from corn, wheat and rice in order to promote the expansion of the product (Morales et al., 2015). As the protein content increases during blending of flours, for instance by adding pulses, the level of expansion in the product declines creating denser structures. Today, pulses are being extruded alone or in-combination with cereal flours for their nutritional properties (Day and Swanson, 2013) and for developing novel food structures with unique texture profiles. For instance, durum wheat semolina (containing vital wheat gluten) was fortified with 10% navy bean flour and pinto protein concentrates to produce spaghetti (Gallegos-Infante et al. 2010). Sensory evaluation of the fortified spaghetti showed acceptable scores with good shelf stability. Faba bean protein concentrates produced by air classification have also been used for the partial replacement of wheat flour in noodle formulations without compromising the color and texture ((Lorenz et al., 1979). In another instance, lentil-corn flour blends were used to create puff snacks with acceptable sensory properties to consumers (Lazou and Krokida (2010). The level of expansion and porosity in these lentil-corn puffs could be controlled by altering the temperature, resident time within the barrel and moisture levels. Extrusion processing has been used to alleviate the hard-to-cook (HTC) defect found in some pulses, and to increase the nutritional value of the bean flours (Martin-Cabrejas et al., 1999; Steinkraus, 2018). Pulses with remarkable nutritional profiles and healthy attributes are truly functional foods. However, direct expanded, RTE, extruded pulses and pulse-based food products are not yet available. Pulse-based healthy food alternatives can secure a sector of the market that is occupied today by snacks that are high in calories, high on the glycemic index and low in protein and dietary fibers (Berrios et al., 2013)

Over the past three decades, extrusion technology had been used effectively to reduce the cooking time of pulses, as well as inactivate undesirable compounds (anti-nutrients), improve the textural, nutritional and sensory characteristics in the final pulse and pulse-based extrudates. Therefore, extrusion processing provides the potential for using pulses and pulse-based mixes in the fabrication of value-added food products and ingredients with high nutritional and economic values.

Pulse flakes

Pulse flakes are only present in the ready- to-eat (RTE) market as niche products, but could be a means of increasing pulse consumption, if the challenges of their structural weakness, "unpleasant" flavours and long cooking times can be overcome. Split desi chickpeas, spilt faba beans, whole *kabuli* and whole desi chickpeas can be processed into flakes. The steps involved in flakes preparation are cooking, partial drying, passing through roller flaker with two different roller gaps and finally drying using a fluidized bed dryer or tray dryer at either. Rolled flakes ranged in thickness from 0.6 - 1.9 mm depending upon the roller gap selected and formulation. Thicker flakes are more robust, but are relatively less robust compared to cereal flakes.

Legumes as functional foods

Legumes contains non-nutrient phytochemicals such as polysaccharides (non-starch polysaccharides (NSP)), phytosterols, saponins, isoflavones, phytoestrogens, phenolic compounds and antioxidants (tocopherols and flavonoids), which promote health and prevent disease.

Isoflavones (Phytoestrogens)

Isoflavones are phytoestrogens that had weak estrogen activity and are present in significant amounts in pulses and legumes. It plays a preventive role against hormone-related diseases such as breast cancer and heart disease. Isoflavones also have natural functions besides estrogen-related activity, for example, antioxidant activity. The key isoflavones are daidzein, genistein, biochinin A, protensin, formononetin, etc. There are reports on the potential of chickpea (bengal gram) for reduction of blood cholesterol and other legumes including soybean. The hypocholesterolemic strength of these legumes is said to be recognized by their isoflavone content. Isoflavone content of pulses is reported to increase after germination. However, the strength of isoflavones decreases after the testicles are cooked. Studies on soy have suggested that the combination of soy protein and isoflavones is needed to produce a low cholesterol effect.

Currently, there is a great interest in soy as a health food because of these biological activities. Isoflavones extracted from soy are sold commercially. Similarly, the combination of soy protein of 6.25 g with 12.5 mg of total (extract) isoflavones, equivalent to 25 g of soy alone, is marketed as a health food (Kutos et al., 2003). Thus, there is considerable scope for future research on Indian pulses for their isoflavones contents, their composition and bioactivity to exploit isoflavone rich pulses as health foods, in the same way that soy has been exploited.

Other phytochemicals

Several other phytochemicals such as saponins, flavons, phenolic acids are said to be present in soybeans and their biological activity has been broadly studied.

Saponins, which are hypocholesterolemic agents, are said to be present in soy in high concentrations. The existence of these phytochemicals in the Indian pulses does not seems to have been studied significantly. The occurrence of these phytochemicals in Indian pulses, their identification, content, chemical nature and their health benefits requires a systematic study. There is also a necessity to learn about the bioavailability of isoflavones and other phytochemicals present in Indian pulses.

Antioxidant properties

As the pulses contain tocopherols, flavonoids and isoflavonoids, all of which can act as antioxidants. Although dry pulses do not contain ascorbic acid, but unripe green pulses contain ascorbic acid which also has antioxidant activity. Both dry legumes and green pulses are a good source of antioxidants. Flavonoids are said to have four to five times more antioxidant activity compared to ascorbic acid. However, the studies on content of flavonoids and their antioxidant potency in Indian pulses are limited.

Factors affecting cooking quality of pulses

Many cooking methods are used for legume food preparations. Some of these are wet heat method and some are dry heat method. Many factors that affect cooking quality of pulses are discussed below:

a. Inherent character: Some varieties are hard – to – cook inherently. Cooking time of whole seed is usually the highest in chick pea followed by pigeon pea, black gram and green gram.

b. Environmental factors: Variation in cooking quality of pulses within and between varieties could be due to location, soil fertility, soil moisture and other environmental factors.

c. Storage condition: Cooking quality is influenced by time, temperature and relatively humidity during storage. In general, cooking time increases with storage time. Moisture content during storage above 10 percent may cause deterioration in the cooking quality.

d. Seed maturity: High temperature at the time of maturing affects the cooking time. Cooking time increases with the increase in seed maturity. The very hard mature seeds take long time to cook.

e. Dehulling: This reduces the cooking time by 50-70 percent and increases digestibility.

f. Soaking: The pulses are soaked in water to improve the cooking quality. Bengal gram cannot be cooked to the desirable consistency without this pretreatment.

g. Salts: Hard – to – cook pulses are soaked in salt solution consisting of 1 percent NaC1 and 0.75 percent NaHCO3 in order to improve the cooking quality. Carbonate or bicarbonate not only act as an alkaline agent and buffer but also acts as a protein dissociating, solubilizing or tenderizing agent. Salts such as tri-sodium phosphate, sodium bicarbonate and ammonium carbonate in small quantities improves cooking quality without appreciably raising the salty taste. These chemicals could be added to cooking water or impregnated or coated on the surface of the *dhal* as a final step in milling.

h. Precooking: The cooking time for precooked lentil seeds is less compared to untreated ones. Precooking is done by cooking, treating with enzymes and dehydrating in controlled conditions.
i. Phytin content: High available phosphorous in the soil contributes to high phytin content in the seed and consequently delay cooking.
j. Calcium and magnesium: Large amounts of insoluble calcium and magnesium pectates are formed in the middle lamella of the cell walls when the seed is high in calcium and magnesium or when the cooking water is high in these elements, which adversely affects the cooking quality.

Conclusion

Pulses have been an integral part of our food systems since time of Vedas. As pulses are a food of global significance owing to their high protein, dietary fibre content and benefits to agriculture. Pulses consumption is below recommended levels in the developed countries and is further declining. Many technologies are being used for pulse consumption, which include traditional methods like fermentation, germination, milling, roasting, parching etc. These traditional home based methods evolved through the time and new mechanized systems and automated technologies were introduced which made the pulse processing business easy. Various novel products have been developed like extruded products, ready to cook products, canned pulses, convenience foods and many more. Mechanisms to ensure access to technology and expertise among local and small-scale food processors should be enhanced. Although, cost might hinder the provision of commercially available starter cultures, delivery of such starter cultures for improved and effective fermentation could be achieved using dried forms of previous fermented products (with viable fermenting organisms), for subsequent use. Most importantly increasing awareness of pulses and consequent fermented products from such crops as sources of functional and health-promoting foods would be the role of government, non-governmental organizations, and other relevant stakeholders within the health and other related sectors. This will to a large extent ensure that developing nations achieve the much-needed and envisaged food and nutrition security.

References

Alavi S., 2004. Impact Report. Available online at http://www.google.com/search?q=Alavi+S.+Impact+Report+2004&ie=utf-8&oe=utf-8&aq=t&rls=org.mozilla:en-US:official&client=firefox-a.

Alonso R., Aguirre A., Marzo, F. (2000). Effects of extrusion and traditional processing methods on antinutrients and in vitro digestibility of protein and starch in faba and kidney beans. *Food Chemistry*, 68, 159-165.

Bede, E. N. (2007). Effect of quenching on cookability of some food legumes. *Food Control, 18*(10), 1161–1164. https://doi.org/10.1016/j.foodcont.2005.03.005.

Benouis, K. (2017). Phytochemicals and bioactive compounds of pulses and their impact on health. *Chem. Int, 3*, 224-229.

Berrios, J. D. J. (2006). Extrusion cooking of legumes: Dry bean flours. *Encyclopedia of Agricultural, Food and Biological Engineering, 1*, 1-8.

Berrios, J. D. J., Ascheri, J. L. R., & Losso, J. N. (2013). Extrusion processing of dry beans and pulses. *Dry Beans and Pulses*, 185-203.

Berrios, J. D. J., Morales, P., Cámara, M., & Sánchez-Mata, M. C. (2010). Carbohydrate composition of raw and extruded pulse flours. *Food Research International, 43*(2), 531-536.

Bhatia, B. S., Ramanathan, L. A.. Prasad, M. S. & Vijayaraghavan, P. K., 1967. Fd. Technol.. Champaigns, 21, 105.

Campos-Vega R., Loarca-Piña G.F., Oomah B.D. (2010). Minor components of pulses and their potential impact on human health. *Food Research International*, 43, 461-482

Chitra U, Vimala V, Singh U, et al. (1995) Variability in phytic acid content and protein digestibility of grain legumes. *Plant Foods for Human Nutrition*, 47, 163-172.

Day, L., & Swanson, B. G. (2013). Functionality of protein-fortified extrudates. *Comprehensive Reviews in Food Science and Food Safety, 12*(5), 546-564.

Dhaliwal, Y. S., & Aggarwal, R. K. (1999). Composition of fat in soybeans as affected by duration of germination and drying temperature. *Journal of Food Science and Technology* (Mysore), 36(3), 266-267.

El-Adawy T.A. (2002). Nutritional composition and antinutritional factors of chickpeas (*Cicer arietinum L.*) undergoing different cooking methods and germination. Plant Foods Hum Nutr., 57(1):83-97. doi:10.1023/a:1013189620528.

Enneking D., Wink M. (2000). Towards the elimination of anti-nutritional factors in grain legumes. In R. Knight (Ed.), Linking Research and Marketing Opportunities for Pulses in the 21st Century (pp. 671-683). Netherlands: Kluwer Academic Publishers.

FAO (2007). *Cereals, Pulses, Legumes and Vegetable Proteins*. Food & Agriculture Org.

Fiave, J. (2018). Instant Cooking Cowpeas. *International Journal of Science and Research (IJSR), 7*(2), 135–141.

Flight I., Clifton P. (2006). Cereal grains and legumes in the prevention of coronary heart disease and stroke: A review of the literature. *European Journal of Clinical Nutrition*, 60, 1145-1159.

Frias, J., Fernandez-Orozco, R., Zielinski, H., Piskula, M., Kozlowska, H., & Vidal-Valverde, C. (2002). Effect of germination on the content of vitamins C and E of lentils. *Polish Journal of Food and Nutrition Sciences, 11*(SPEC; ISS 1), 76-78.

Frias, J., Martinez-Villaluenga, C., & Peñas, E. (Eds.). (2016). *Fermented Foods in Health and Disease Prevention*. Academic Press.

Gallegos-Infante; José, RG; Nuria, GLR; Ochoa, A; Corzo, N; Bello P L; Medina,T L; Peralta, A L. (2004). Quality of spaghetti pasta containing Mexican common bean flour (*Phaseolus vulgaris* L.). Food Chemistry, 119, doi- 10.1016/j.foodchem.2009.09.040.

Gonçalves A., Goufo P., Barros A., Domínguez-Perles R., Trindade H., Rosa E.A.S., Ferreira L, Rodriguesa M. (2016). Cowpea (*Vigna unguiculata* L.Walp), a renewed multipurpose crop for a more sustainable agri-food system: nutritional advantages and constraints. *Journal of the Science and Food Agriculture*, 96, 2941-2951.

Goyal, R. K. Wanjari, O. D. Ilyas, S. M. Vishwakarma, R. K. Manikantan, M. R. and Mridula, D. (2005). Pulse Milling Technologies. Technical Bulletin, CIPHET/Pub/2005. Central Institute of Post-Harvest Engineering and Technology, Ludhiana, India.

Hefnawy T.H., (2011). Effect of processing methods on nutritional composition and anti-nutritional factors in lentils (*Lens culinaris*). *Annals of Agricultural Science*, 56(2), 57-61.

Hiregoudar, S. Sandeep, T. N. Nidoni, U. Shrestha, B. and Meda, V. (2014). Studies on dhal recovery from pre-treated pigeon pea (*Cajanus cajan* L.) cultivars. *J. Food Sci. Technol.* 51(5):922–928.

Iwuoha, C. I., & Eke, O. S. (1996). Nigerian indigenous fermented foods: their traditional process operation, inherent problems, improvements and current status. *Food Research International*, 29(5-6), 527-540.

Jayaraman K 1984. Development of instant foods and emergency-survival rations for service use. *Defence Science Journal*, vol. 34, issue 2 (1984) pp. 161-172.

Jayaraman, K. S., Gopinathan, V. K. & Ramanathan, L.A., J. *Fd. Technol.*, 15 (1980), 217.

Jayaraman, K. S., Gopinathan, V. K., & Ramanathan, L. A. (1980). Development of quick-cooking dehydrated pulses by high temperature short time pneumatic drying. *International Journal of Food Science & Technology, 15*(2), 217–226. https://doi.org/10.1111/j.1365-2621.1980.tb00933.x

Jayaraman, K. S., Gopinathan, V. K., Pitchamuthu, P. & Vijayaraghavan, P. K., (1981) 'Preparation of quick-cooking dehydrated vegetables and grams by high temperature short tlme drying', Paper presented at the Second Indian Convention of Food Scientists and Technologists (AFST, Mysore), Feb., Abstracts of papers, p. 69,.

Joyner, J. J. and Yadav, B. K. (2015a). Microwave assisted dehulling of black gram (Vigna Mungo L.). J. *Food Sci. Technol.* 52(4):2003–2012.

Joyner, J. J. and Yadav, B. K. (2015b). Optimization of continuous hydrothermal treatment for improving the dehulling of black gram (Vigna mungo L). *J. Food Sci. Technol.* 52(12):7817–7827.

Khattab R.Y., Arntfield S.D. (2009). Nutritional quality of legume seeds as affected by some physical treatments 2. Antinutritional factors. *LWT-Food Science and Technology*, 42, 1113-1118.

Köksel, H., Sivri, D., Scanlon, M. G., & Bushuk, W. (1998). Comparison of physical properties of raw and roasted chickpeas (leblebi). *Food Research International, 31*(9), 659-665.

Kutoš, T., Golob, T., Kač, M., & Plestenjak, A. (2003). Dietary fibre content of dry and processed beans. *Food Chemistry, 80*(2), 231-235.

Lal, R. R. and Verma, P. (2007). Post-Harvest Management of Pulses. Indian Institute of Pulse Research, Kanpur, India.

Lazou, A., Krokida, M., & Tzia, C. (2010). Sensory properties and acceptability of corn and lentil extruded puffs. *Journal of Sensory Studies, 25*(6), 838-860.

Lorenz, K., Dilsaver, W., &Wolt, M. (1979). Fababean flour and proteinconcentrate in baked goods and in pasta products. *Bakers Digest*, 39, 45–51.

Luo, Y., Gu, Z., Han, Y., & Chen, Z. (2009). The impact of processing on phytic acid, in vitro soluble iron and Phy/Fe molar ratio of faba bean (*Vicia faba* L.). *Journal of the Science of Food and Agriculture, 89*(5), 861-866.

Martín-Cabrejas, M. A., Jaime, L., Karanja, C., Downie, A. J., Parker, M. L., Lopez-Andreu, F. J., ... & Waldron, K. W. (1999). Modifications to physicochemical and nutritional properties of hard-to-cook beans (*Phaseolus vulgaris* L.) by extrusion cooking. *Journal of Agricultural and Food Chemistry, 47*(3), 1174-1182.

Morales, P., Cebadera-Miranda, L., Cámara, R. M., Reis, F. S., Barros, L., Berrios, J. D. J., & Cámara, M. (2015). Lentil flour formulations to develop new snack-type products by extrusion processing: Phytochemicals and antioxidant capacity. *Journal of Functional Foods, 19*, 537-544.

Mubarak, A.E. (2005). Nutritional composition and antinutritional factors of mung bean seeds (*Phaseolus aureus*) as affected by some home traditional processes. *Food Chemistry*, 89, 489-495.

Nagaraju, V. D., & Manohar, B. (2000). Rheology and particle size changes during Idli fermentation. *Journal of Food Engineering*, 43(3), 167-171.

Narasimha, H. V. Ramakrishnaiah, N. and Pratape, V. M. (2003). Milling of pulses. In: Handbook of Post Harvest Technology, pp. 427–454. Chakraverty, A. Mujumdar, A. S. Raghavan, G. S. V. and Ramaswamy, H. S., Eds., Marcel Dekker, Inc., New York, USA.

Nayak L.K. and D.V.K. Samuel (2015). Process optimization for instant pigeon-pea dal using NaCl (Sodium chloride) pretreatment. Agric. Sci. Digest., 35 (2) 2015: 126-129

Oomah B.D., Blanchard C., Balasubramanian P. (2008). Phytic acid, phytase, minerals, and antioxidant activity in Canadian dry bean (*Phaseolus vulgaris* L.) cultivars. *Journal of Agricultural and Food Chemistry*, 56, 11312-11319.

Osorio-Díaz, P., Bello-Pérez, L. A., Sáyago-Ayerdi, S. G., Benítez-Reyes, M. D. P., Tovar, J., & Paredes-López, O. (2003). Effect of processing and storage time on *in vitro* digestibility and resistant starch content of two bean (*Phaseolus vulgaris* L) varieties. *Journal of the Science of Food and Agriculture*, 83(12), 1283-1288.

Patki PE, Arya SS (1994) Studies on development and storage stability of instant dhals. *Ind Food Pckr* 48:31–39.

Patwardhan, V. N. (1962). Pulses and beans in human nutrition. *American Journal of Clinical Nutrition, 11*, 12-30.

Perez-Hidalgo, M. A., Guerra-Hernández, E., & Garcıa-Villanova, B. (1997). Dietary fiber in three raw legumes and processing effect on chick peas by an enzymatic-gravimetric method. *Journal of Food Composition and Analysis*, 10(1), 66-72.

Pierson, M. D., Reddy, N. R., & Odunfa, S. A. (1986). Other legume-based fermented foods. agris.fao.org.

Ramakrishnaiah, N., Pratape, V. M., and Narasimha, H. V. (1993). National Survey of Pulse Milling Industry in India including Rural Processing: A Status Report. Central Food Technological Research Institute, Mysore, India.

Reddy N.R., Pierson M.D., Sathe S.K., Salunkhe, D.K. (1985). Dry bean tannins: A review of nutritional implications. *Journal of the American Oil Chemists Society*, 62 (3): 541-549.

Rehman, Z. U., Salariya, A. M., & Zafar, S. I. (2001). Effect of processing on available carbohydrate content and starch digestibility of kidney beans (*Phaseolus vulgaris* L.). *Food Chemistry, 73*(3), 351-355.

Ridout, M. J., Gunning, A. P., Parker, M. L., Wilson, R. H., & Morris, V. J. (2002). Using AFM to image the internal structure of starch granules. *Carbohydrate Polymers*, 50(2), 123-132.

Sands, D. C., & Hankin, L. (1974). Selecting lysine-excreting mutants of lactobacilli for use in food and feed enrichment. *Appl. Environ. Microbiol., 28*(3), 523-524.

Sathe, S. K. (2002). Dry bean protein functionality. *Critical Reviews in Biotechnology*, 22(2), 175-223.

Scheule. (2016). Schule-Technical Literature on Verti Sheller and Splitter for Pulses. FH Schule Muhlenbau GmbH, Dieselstrasse 5, D-21465, Reinbek, Hamburg, Germany.

Sharma, Monika; Yadav, Deep N; Mridula, D; Gupta, RK. (2015) Protein Enriched Multigrain Expanded Snack: Optimization of Extrusion Variables. *Proceedings of the National Academy of Sciences, India Section B: Biological Sciences*. DOI 10.1007/s40011-015-0546-5

Shi L. (2015). A quantitative assessment of the anti-nutritional properties of canadian pulses. M.Sc. Thesis, Department of Food Science, University of Manitoba, Winnipeg, Manitoba.

Shimelis, E. A., & Rakshit, S. K. (2007). Effect of processing on antinutrients and *in vitro* protein digestibility of kidney bean (*Phaseolus vulgaris* L.) varieties grown in East Africa. *Food chemistry, 103*(1), 161-172.

Shruti Sethi & D. V. K. Samuel & Islam Khan (2011). Development and quality evaluation of quick cooking dhal—A convenience product. *J Food Sci Technol* DOI 10.1007/s13197-011-0534-6.

Singh U (2007) Processing and food uses of grain legumes. In: Katkar BS (ed) Food science and technology. Daya Publishing House, Delhi, pp 70–79.

Singh U. (1988). Anti-nutritional factors of chickpea and pigeon pea and their removal by processing. *Plant Foods for Human Nutrition*, 38, 251-261.

Singh, A. and Ilyas, S. M. (1987). Energy audit of pulse milling technologies in India. Paper presented in National Energy Conference on Energy in Production Agriculture and Food Processing, Punjab Agricultural University, Ludhiana, India, October 30-31, 1987.

Singh, J., & Basu, P. S. (2012). Non-nutritive bioactive compounds in pulses and their impact on human health: An overview.

Singh, U. (1993). Protein quality of pigeon pea (*Cajanus cajan* L.) as influenced by seed polyphenols and cooking process. Plant *Food Human Nutr.* 43:171–179.

Singh, U. (1995). Methods for dehulling of pulses: A critical appraisal. *J. Food Sci. Technol.* 32(2):81–93.

Singh, U., Santosa, B. A. S., and Rao P. V. (1992). Effect of dehulling methods and physical characteristics of grain on dhal yield of pigeon pea (*Cajanus cajan* L.) genotypes. *J. Food Sci. Technol.* 29: 350-353.

Soetan K.O., Oyewole O. E. (2009). The need for adequate processing to reduce the antinutritional factors in plants used as human foods and animal feeds: A review. *African Journal of Food Science*, 3(9), 223-232.

Sokhansanj, S. and Patil, R. T. (2003). Dehulling and splitting pulses. In: Handbook of Post Harvest Technology, pp. 397–426. Chakraverty, A. Mujumdar, A. S. Raghavan, G. S. V. and Ramaswamy, H. S., Eds., Marcel Dekker, Inc., New York, USA.

Solomon S.G., Victor T.O., Oda S.O. (2017). Nutritional value of toasted pigeon pea, Cajanus cajan seed and its utilization in the diet of *Clarias gariepinus* (Burchell, 1822) fingerlings. Aquaculture Reports 7, 34-39. http://dx.doi.org/10.1016/j.aqrep.2017.05.005.

Sosulski, K. Sosulski, F. W. and Coxworth, E. (1988). Carbohydrates hydrolysis of canola to enhance extraction with hexane. J. Am. Oil Chem. Soc. 65:357–361.

Sreerama, Y. N. Sashikala, V. B. and Pratape, V. M. (2009). Effect of enzyme pre-dehulling treatments on dehulling and cooking properties of legumes. *J. Food Eng.* 92:389–395.

Srilakshmi, B. (2019). In Book: Food Science, New Age International (P) Ltd., Publishers, New Delhi.

Srivastava, R. P., & Srivastava, G. K. (2006). Accumulation of β-N-Oxalyl Amino Alanine in Lathyrus during Podding and Maturation. *Indian Journal of Agricultural Biochemistry*, *19*(1), 39-41.

Srivastava, V. Mishra, D. P. Chand, C. Gupta, R. K. and Singh, B. P. N. (1988). Influence of soaking on various biochemical changes and dehusking efficiency in pigeon pea (*Cajanus cajan* L.) seeds. *J. Food Sci. Technol.* 25:267–271.

Steinkraus, K. (2018). *Handbook of Indigenous Fermented Foods, Revised and Expanded.* CRC Press.

Suneja Y., Kaur S., Gupta A.K., Kaur N. (2011). Levels of nutritional constituents and antinutritional factors in black gram (*Vigna mungo* L. Hepper). *Food Research International*, 44, 621-628.

Tiwari B.K., Gowen A., McKenna B. (2011). Pulse Foods: Processing, Quality and Nutraceutical Applications. Elsevier Inc.

Tiwari, B. K. Jaganmohan, R. and Vasan, B. S. (2007). Effect of heat processing on milling of black gram and its end product quality. *J. Food Eng.* 78:356–360.

Vandenberg, A. (2009). Postharvest processing and value addition. In: The Lentil: Botany, Production and Uses, pp. 391–407. Erskine, W. Muehlbauer, F. J. Sarkar, A. and Sharma, B., Eds., CAB International, Wallingford, UK.

Vasishtha, H., & Srivastava, R. P. (2013). Effect of soaking and cooking on dietary fibre components of different type of chickpea genotypes. *Journal of Food Science and Technology*, *50*(3), 579-584.

Venter CS, Eyssen E. (2001). More Legumes for Overall Better Health. Dry bean Producers Organisation, Pretoria. *SAJCN (Supplement);* 14(3):S32-S38.

Verma, P. Saxena, R. P. Sarkar, B. C. and Omre, P. K. (1993). Enzymatic pre-treatment of pigeon pea (*Cajanus cajan*) grain and its interaction with milling. *J. Food Sci. Technol.* 30:368–370.

Vidal-Valverde, C., Frias, J., Sierra, I., Blazquez, I., Lambein, F., & Kuo, Y. H. (2002). New functional legume foods by germination: effect on the nutritive value of beans, lentils and peas. *European Food Research and Technology, 215*(6), 472-477.

Vishwakarma RK, Shivhare US, Gupta RK, Yadav DN, Jaiswal A, Prasad P (2017). Status of pulse milling processes and technologies: A review. Critical Reviews in Food Science and Nutrition, 1-14, DOI:10.1080/10408398.2016.1274956.

Wang, N. (2005). Optimization of a laboratory dehulling process for lentils (*Lens culinaris*). *Cereal Chem.* 82(6):671–676.

Wang, N., & Toews, R. (2011). Certain physicochemical and functional properties of fibre fractions from pulses. *Food Research International*, 44(8), 2515-2523.

Williams, M., Tian, Y., Jones, D. S., & Andrews, G. P. (2010). Hot-melt extrusion technology: optimizing drug delivery. *European Journal of Parenteral Sciences and Pharmaceutical Sciences*, 15(2), 61.

Wood, J. A. and Malcolmson, L. J. (2011). Pulse milling technologies. In: Pulse Foods: Processing, Quality and Nutraceutical Applications, pp. 193–221. Tiwari, B. Gowen, A. and McKenna, B., Eds., Academic Press, UK.

Yadav D.N., Anand T., Navnidhi and Singh A.K. (2013). Co-extrusion of pearl millet-whey protein concentrates for expanded snacks. *International Journal of Food Science and Technology*. DOI:10.1111/ijfs.12373.

Yasmin A., Zeb A., Khalil A.W., Paracha G.M., Khattak A B. (2008). Effect of processing on anti-nutritional factors of red kidney bean (*Phaseolus vulgaris*) grains. *Food and Bioprocess Technology*, 1, 415-419.

6

Recent Trends in Technologies of Cereal Based Food Products

*Ritu Sindhu[1] and B. S. Khatkar[2]**

[1]Centre of Food Science and Technology, CCSHAU, Hisar, Haryana

[2]Department of Food Technology, Guru Jambheshwar University of Science and Technology, Hisar, Haryana

Introduction

Cereals and cereal products comprise a long history of consumption by humans. Cereals are the important source of nutrients and non nutritive health beneficial components and for that reason cereal based food products are consumed as staple food in human diet in developed as well as developing countries. The cereal grains mainly cultivated and consumed as food include rice, maize, wheat, sorghum rye, barley, oats, and millet. Products obtained from cereal grains are flours, grits, semolina, breakfast cereals, extruded product like pasta, puffs and flakes, and bakery products such as bread, biscuits, cakes and tortillas. Moreover, cereals or their components are used as ingredients in the preparation of various food products such as beverages infant foods coatings and processed meats. Initially cereals are consumed as whole or with minimal processing like boiling, grinding and roasting. However, processing like salting, drying, steeping and fermentation of cereals for preparation of cereals based beverages and foods has been practiced since the initiation of cereals consumption as food by human. Increasing consumer awareness and commercialization of cereal products have created interest of food manufacturers in utilization of novel and more efficient technologies with optimized conditions, development of convenient and more health beneficial foods with satisfactory sensory properties and wide variety of foods to ease the choice of consumers. Market full of diverse range of snack foods, breakfast cereals, ready to eat and ready to cook foods, and novel functional foods with variation in appearance, texture and flavor is an indication of evolution of food industry from basic to complicated form focused on efficient and automated operations. In ancient times, fermentation of food was done with the aim of enhancement of shelf life and palatability without clearly knowing the role of microorganisms. Modern fermentation technology uses various operations and specific strains of microorganisms for conversion of bulky and perishable materials into novel foods with more pleasing flavour, enhanced functional properties and extended shelf life. In addition to basic bread, new varieties of bread developed with improved functionality like sourdough bread, gluten free breads and high fibre breads is an example of innovation in cereal fermented foods indicating modification in ingredients, microbial culture and processing technologies as compare to traditional fermentation process. The word "extrudate" is derived from the Latin word "ex" meaning 'out' and "trudere" meaning 'to thrust'. Extrusion technology is turning out to be very popular in food industry due to its versatility, usefulness and cost efficiency. It is well-organized state-of-art technology and plays important role in various food industrial processes such as cooking, mixing, conveying, shearing and shaping of end products. Extrusion cooking is high temperature short time (HTST) with high pressure process, is especially used for preparation of cereal based snacks, fat replacers, breakfast cereals, infant formula, gluten free foods, starch and protein rich cereals food products. A wider range of ready to eat food products with different texture, size and shape is produced by fast and easy extrusion cooking. In epidemiological studies it is stated that whole grain consumption is more health beneficial than refined grains and the reason behind it is that dietary fibre and phytochemicals located in germ and seed coat are generally removed during milling or polishing. Various technologies are developed by food scientists and technologists to preserves the bioactive compounds during processing and or to supplement processed food products with the bioactive compounds during food preparation to maintain the bioactivity and functionality of final products. Utilisation of whole grain flours, composite flours, supplementation of cereal foods with ingredients from fruits, vegetables, plant extracts and legumes, bio-fortification of crops, implementation of genetic engineering to produce crops rich in bioactive compounds are some technologies which are being used in modern food industry to develop value added cereal food products.

Baking technology

The term bakery products covers a wide variety of foods having one common thread that in their recipes wheat flour in a significant ratio with other ingredients is the important component. Basic ingredients of bakery products have been flour and water since ancient times while in today's bakery products various ingredients including sweeteners, fats, salts, leavening agents and flavouring agents are added. The evolution of bakery products from the basic simple forms to the present range with specific features and functionalities using complicated mechanized technologies has been going hand-in-hand with the advancement in other interrelated areas of science and technology. Bread is the most popular bakery product while other bakery products like cookies, biscuits, pastries and cakes are also not lagging far behind. The combinations of ratio of ingredients and processing conditions are the main basis for variation in bakery products.

Ingredients

Major ingredients of baked foods are flour, yeast, sugar, water, egg and dairy products. Flour is the key ingredient in every baked product and plays important role in nutritional value, binding of other ingredients, organoleptic and textural properties of final product. In earlier times only whole wheat flour was used in bakery products but in present times millers work together with bakers to produce the right flour for the baker's products. A wide range of flours from wheat and non-wheat sources with specific composition suitable for preparation of different bakery products is available in market. Still wheat flour alone or in composite form is used as main ingredient in preparation of mostly bakery products. In addition to nutrients, wheat flour contains gluten (insoluble protein) which is responsible for rheological properties of dough and texture as well as volume of end products. The formation of bread loaf from raw ingredients is generally expressed as a process of transformation of foam to sponge. The characteristic structure of bread is attributed to the separation of gas babbles in foam by disconnecting them from each other by gluten network. However, products like cake and sponge have cellular structure but these are prepared from low gluten flours and foam formation and stabilization mechanisms are different from bread. In these products fats, proteins and emulsifiers are mainly responsible for spongy structure. Preparation of products such as biscuits, cookies, and pastries requires low gluten flour due to low water level in their recipes and gluten rich flour can lead to quality defects in end product. In pastry like products, strong gluten network is required in dough to make it suitable for formation of alternate layers of fat and dough to provide typical flaky texture to the final product. Starch is another main content of the flour that gelatinized during baking and strengthens the baked product.

Leavening agents including yeasts, chemicals and some microorganisms are used to produce carbon dioxide in dough for expansion or fermentation. Expansion of dough is important for volume and texture of end product in bakery. Each leavening agent has its capacities and limitations that make it suitable for particular application. Generally chemical leavening agents are used for fast production on carbon dioxide. Saccharomyces Cerevisiae is the known as baker's yeast and used in bakery industry as raising agent. Enzymes present in the yeast hydrolyze the starch to glucose and fructose in the flour. In bakery industry the yeast is used in mainly three forms- fresh yeast or compressed yeast, dry yeast and instant yeast or powered dried yeast. Dough Improvers like potassium bromated, azodicarbonamide, sodium metabisulfite and l-cysteine are used to improve rheological properties of baked products. In some countries only oxidising agents are allowed to be used and redox materials are prohibited. Now a day's some selected enzymes are used as substitute of traditional redox materials for the production of baked foods. A wide range of sweeteners are used to provide desirable taste, appearance, texture and flavour in the bakery products. Sucrose is the most commonly used sweetener but consumers interest has been increasing in low calorie products since last few decades. Along with the sweetener, sucrose acts as preservative, as softness, color, and texture and flavour enhancer, as fermentation aid and as bulking agent in foods. Therefore, replacing the sucrose with suitable substitute is a huge challenge in bakery industry because of evident losses in texture and sensory properties of the final food due the sucrose reduction or substitution.

Fats and lipids are important ingredients in bakery products influencing handling properties and shelf life of dough. Shortenings impart tenderness, mouthfeel, flavour and desirable texture to the final product. Liquid portion of shortening contribute to the mouthfeel and lubricity while solid portion is responsible for dough structure. Selection of any shortening for any particular application depends on oxidative stability, crystal structure and ratio of liquid to solid phase in shortening. Egg is used in bakery products as foaming, emulsifying, and gel network agents which provide unique organoleptic and textural properties to the product. Various dairy products like milk, milk powder, and whey protein and milk fats are used in development of bakery products to improve the nutritional value of products as well as to alter rheological properties of dough. Sometime dairy products are used as substitute of other bakery ingredients like fat and egg.

Processing

Mixing is the first and most important step in the baking process. It includes mixing of all ingredients properly and uniformly to develop a homogenous dough. It is a critical process for quality of dough as if mixing is not done properly it will result in low quality dough that cannot be possible to correct it afterwards. Quality of dough depends on ingredients and mixing process. Different methods are used to develop dough for different bakery products. For instance, mixing and kneading of all the ingredients is done in single step in straight dough method while in sponge dough method dough is developed in two steps. In first step yeast, sugar and the part of the flour is mixed in liquid to make sponge. In second step, other ingredients are added to the bubbly and light sponge. This method is lengthier than straight dough method. Batter method is similar to straight dough method excepting kneading step. Mixing time depends on the method used, desired volume, water absorption capacity of flour, other ingredients and mixer used. Over-mixed and under-mixed doughs have poor volume and texture. During kneading gluten is developed; high pressure at initial stage of kneading produce sticky dough while high pressure at the end of the process can break the gluten strands. Various types of mixers are used for dough making and the speed for mixing and kneading depends on the desired quality of dough.

After development of dough it is allowed to ferment. During fermentation yeasts act on sugars and produce alcohol and carbon dioxide gas. Production of gas causes the expansion of dough and volume of dough becomes almost double in size. During fermentation, acidity of dough increases and also gas retention capacity of dough increases due to some enzymatic activities. Fermentation also alters the starch digestibility and availability of minerals and vitamins. Properly fermented dough is smooth and elastic. The under-fermented dough has coarse texture and lesser rise in volume while over-fermented dough is sticky. Steady fermentation is done using starter cultures and monitoring the dough rising process through non-destructive tests like digital video cameras, magnetic resonance microscopy, X-ray tomography, and acoustic waves. Fermented dough is deflated to expel the gases formed during the fermentation and this treatment is called punching. Then the dough is cut in into small pieces and rolled in round shaped balls. Next step is secondary fermentation or proofing where volume of dough rises due to gas production. Proofing is done to achieve the final volume and develop the gluten which was destroyed during punching. The important factors in the process of baking are time period and temperature of baking which depend on size of the loaf and dough type. During baking instantaneous transfer of heat and mass occurs in the dough causes following changes-

- Initially volume of loaf rise during baking that is called oven spring
- Vaporization of gases on rise of temperature
- Killing of yeasts
- Gelatinization of starch
- Protein coagulation
- Development of porous structure with interconnected pores
- Browning and development of crust

Mainly two baking techniques are used: high temperature short time (HTST) and low temperature long time (LTLT). In high temperature short time technique baking is too fast and crust will be good while low temperature long time baking produces product of thicker rigid skin without browning. After the baking, product is cooled to room temperature as soon as possible using refrigerator or air circulation or evaporation of water on the surface. Shelf life of baked products is small as immediately after the baking process quality of baked products starts to deteriorate by staling, flavour and texture loss. These products are packed generally in plastic materials having good mechanical strength as well as good barrier properties for moisture, oxygen and light. Proper optimization, monitoring and control of process from mixing to packing is required because there is interaction between all processing steps and modification at one processing step may alter the other step. For instance, rising the temperature during proofing will alter the development of dough volume and consecutively baking process due to change in heat mass transfer. Further complications in optimizing the baking process are the timings from mixing to the baking and assurance that modification or alteration made to any step is not compromised by another. These technical hitches create interesting confronts for in the area of baking process optimization.

Recent advances

Baked products are calorie rich while consumer's demand is increasing towards low calorie foods. Usually flour and sweeteners are the major components of sweet baked products and substituting any one or both is a challenge because both the ingredients play vital role in quality of end products from technological and nutritional point of view. Reduction of sucrose level in the development of low sugar baked food can cause noticeable changes in textural, rheological and sensory properties of products. Any newly developed healthy food should also be enjoyable and satisfying at eating for its acceptance. Work is in progress for searching and application of new non or low calorie sweeteners suitable for substitution of sucrose in bakery industry. Honey is sweeter than sucrose and can be used as a sweetener in baked products. It has good water holding capacity that contributes to the texture of end product by delaying the dryness and brittleness; glucose and fructose which make it a good substrate for yeast growth and reducing sugars provide brown color to the baked product. Honey, a natural sweeten, is used in baked products also due to its antioxidant and antimicrobial properties, flavour, color and texture contribution to the product and its gastrointestinal health benefits.

Polyols are other versatile ingredients which can be used as sweeteners, bulking agents, emulsifiers, stabilizers, humectants, thickeners, glazing agents, and anti-caking agents in formulation of various food products. The most widely used polyols are sorbitol, mannitol, erythritol, maltitol, isomalt, lactitol, xylitol, and hydrogenated starch hydrolyzates. In bakery industry polyols can be used as substitute of sucrose having sweet taste, moisture holding capacity and ability to inhibit sugar crystallization. Polyols provide a positive influence on taste, appearance, textural and nutritional properties and shelf-life of baked products (Patton and O'Brien Nabors, 2011). Interest in polyols is increasing also due to their remarkable health benefits like non-carcinogenic, low-glycemic (diabetes and cardiovascular disease), low-insulinemic, low-digestible (dietary fibre), and osmotic (laxative effect and purifying carbohydrates). Polyols act as dietary fibres in intestine as they are not easily digested but ferment easily. Polyols are stable to heat and pH during processing or cooking as they do not undergo non-enzymatic browning due to absence of reducing groups. Moreover, in view of the fact that polyols cause no hazard to consumer health, no ADI value has been fixed for its use. Negative association of cardiovascular diseases with consumption of lipids containing trans and saturated fat has created in interest in developing new semi-plastic structures with absence of undesirable hydrogenated or saturated fats for formulation of healthy foods. As shortenings are important in food processing from nutritional and technological aspects and the new alternative of fats have to fulfil the functionality of traditional lipids containing saturated and trans fats. Monogoyceride gel containing only 55% fat and very low level of saturated fat fraction has been developed by encapsulation technique and utilized in bakery products. This novel structured monoglyceride emulsion showed similar functionality to that of commercial shortening like soy shortening and other liquid oils determined by various instruments like visco-amylograph, farinograph and gluten peak tester (Huschka et al, 2011).

Generally unwanted oil migration takes place in cookies prepared by using low saturated fat shortening. Leakage of oil from low saturated shortenings can be reduced by using organo-gels like ethyl cellulose (EC). Addition of ethyl cellulose in cookies prepared using shortening with low saturated fat showed acceptable functionality with lower content of total fat and saturated fat (Stortz et al, 2012). Although development of baked products by completely removal of fats is not possible but partial replacement by some carbohydrates is possible. Zoulias et al (2002) reported harder and brittle cookies prepared by partial replacement of fat upto 50% using polydextrose, maltodextrin and inulin. Maltodextrin as shortening replacer increased dough hardness and appreciably improved texture of cookies (Sudha et al, 2007). Fat replacement upto 20% by inulin produced cookies with additional health benefits and textural quality similar to control cookies (Rodriguez-Garcia et al, 2012). Various dairy products used in bakery industry have their significant nutritional and functional importance. Whey is used at different levels in bakery products like fermented products (6%), biscuits (10%) and cakes (20%). In addition to being a source of minerals and protein, whey reduces the bread staling and improves organoleptic properties of bread (Bilgin et al, 2006). Divya and Rao (2010) proposed higher dose of yeast or higher temperature during fermentation as whey may decrease the rate of fermentation. Dairy proteins including casein, whey proteins and heat treated whey proteins were reported to improve the textural properties of bread prepared with frozen dough (Asghar et al, 2009). Gluten free breads are generally prepared using maize or potato starches and it required addition of protein in their formulation to improve nutritional value. Nunes et al (2009) reported that whey protein addition in gluten-free breads raised bread volume and reduced hardness during storage while sodium caseinate showed the opposite effect. The action of gluten can be mimicked by structured protein obtained from whey protein and locust bean gum in the development of gluten free bread (Van Riemsdijk et al, 2011). Egg is an essential ingredient in for the preparation of good quality angel food cakes. Owing the good emulsifying properties protein isolates and concentrates obtained from whey act as suitable substitute of egg for angel cakes preparation (Abu-Ghoush et al, 2010).

Baker's yeast is an important functional ingredient of various fermented products based on cereals. The major function of yeast is production of carbon dioxide gas by fermenting the sugars. The carbon dioxide gas released during fermentation increases loaf volume and helps in development of crumb structure of bread. Apart from these functionalities, baker's yeast can be used as source of nutrients like proteins or amino acids, vitamins and as energy and immune system booster in the preparation of health products or nutraceuticals (Mazo et al, 2007). Fermentation of dough by lactic acid bacterial instead of traditional baker's yeast for preparation of bread is new trend in baking industry. Bread developed by lactic acid fermentation is known as "Sourdough bread" and is becoming very popular due to its unique flavour. In development of sourdough bread the principal lactic acid bacteria varied according to the type of flour and the region of bread consumption. Mostly used lactic acid bacteria for sourdough include *L. plantarum, L. curvatus, L. paracasei, L. sanfranciscensis, L. pentosus, L. paraplantarum, L. sakei, L. brevis, Pediococcus pentosaceus, Leuconostoc mesenteroides, Leuconostoc citreum, Weissella cibaria and Lactococcus lactis* (Robert et al, 2009). Demand of yeasts is increasing day by day and traditional sources of yeasts like molasses make yeast propagation process costly. Sometime yeasts have to undergo harsh conditions during cooking or processing which can cause stress on yeasts. Therefore, to overcome these problems advance biotechnological work is going on for the development of specialized yeasts with new and optimised properties at lower cost and with more stress protective response.

Extrusion technology

Extrusion is the process in which shaping of a plastic or dough-like material is done by forcing it using piston or screw through a restriction or die. Therefore, food extruder are the devices mainly used to accelerate the shaping and restructuring of raw material to form ready to cook or ready to eat products with low cost and high productivity. Interest in extrusion-technology is increasing worldwide in the food industry due to its efficient functions like conveying, mixing, shearing, separation, heating or cooling, shaping, co-extrusion, venting volatiles and moisture, flavour generation, encapsulation and sterilization. The quality of extruded end product depends on various parameters such as ingredients, temperature of barrel, moisture content of feed, speed of screw, extruded type and screw configuration. There are two forms of extrusion – cold extrusion process (below 70°C) used in production of pasta and noodle like products and hot extrusion process (above 70°C) used for production of puffed or expanded food products. During extrusion cooking or hot extrusion process the cooking temperature generally falls in the range of 100 – 170°C for a residence period of 20-40 seconds.

Thus, the extrusion cooking is known as high temperature short time (HTST) cooking process. An extrusion technology presents the opportunity to develop a wide range of food products only by minute or slight alteration in ingredient and or processing conditions on extruder during extrusion. As compared to traditional methods of cooking the extrusion method is more energy efficient and provides the opportunity for production of novel and creative breakfast cereals and snack foods at lower cost (as shown in Table 1). In food industry extruders have become the benchmark working equipment for preparation of breakfast cereals and snack food globally.

Table 1: Advantages of extrusion cooking process

Feature	Advantages
Adaptability	Production of wide range of products is practicable just by minor alteration in raw material and extrusion cooking conditions. It is amazingly flexible technology for developing new products of consumers demand
Product features	Feasible production of products with various shapes, sizes, texture and appearances
Energy efficiency	Extrusion cooking is done at comparatively low moisture therefore requisite lesser re-drying
Low cost	Extrusion requires lower capital cost and lesser space than traditional processes; and saves manpower as well as raw materials expenditure requires

In extrusion technology three important factors for developing new products or making variations in products are -ingredients, type of extruder and processing conditions. Depending on the suitability different designs of extruder are used in food and non food industries. For instance, some extruders are designed to transmit the ingredients while other are designed to do mixing and kneading of ingredients. In extrusion cooking extruder are intended to provide mechanical and thermal energy to the raw materials to get end product of desired physico-chemical properties.

Recent advances

Initially singly screw was used in 1935 for plasticising thermoplastic materials and application of twin-screw extruders started in the mid 1930s in food industry for food products formulation. Later on, single-screw extruders were appeared into utilisation in pasta industry for development of spaghetti and macaroni-type products (Moscicki and Zuilichem, 2011). In snacks food industry mainly piston extruders, roller-type extruders, and screw extruders are used. In food industry mainly two types of screw extruder are used: Single-screw extruder and Twin-screw extruder. Extruders with more than two screws are used in plastic or non-food industries.

In modern time, consumers have turn out to be more health conscious and prefer healthier and more nutritious breakfast and snack products than which were previously available. Therefore, the demand for a convenient, ready to eat products with low fat and sugar and fortified with fibres, minerals and vitamins is increasing.

However it's a challenge to produce a healthier snack or ready to eat food with enhanced nutritional value and appealing appearance due to problems associated with ingredient formulation and processing constraints like more hardness, lesser expansion and lower sensory score. Here extrusion process offers a satisfactory solution to these hurdles by providing an efficient process for production of wide variety of ready to eat and ready to cook food products with enhanced nutritional and sensory properties. The qualitative properties of extruded product are important factors for its acceptance by consumers. The extent of alteration in structure and texture of the products during extrusion actually determined the qualitative properties of end product. The level of modification or the properties of final product depend on different factors such as- ingredient properties like composition and moisture content; processing parameters like speed of screw and barrel temperature; and configuration of extrusion equipment including screw profile, barrel size and size and shape of the die (Figure 1). It was also reported that twin screw extruder had better ingredient mixing ability and more pumping efficiency while higher capital as well as running cost and more mechanical complexity were the limitations as compared to singly screw extruders (Martha et al, 2017).

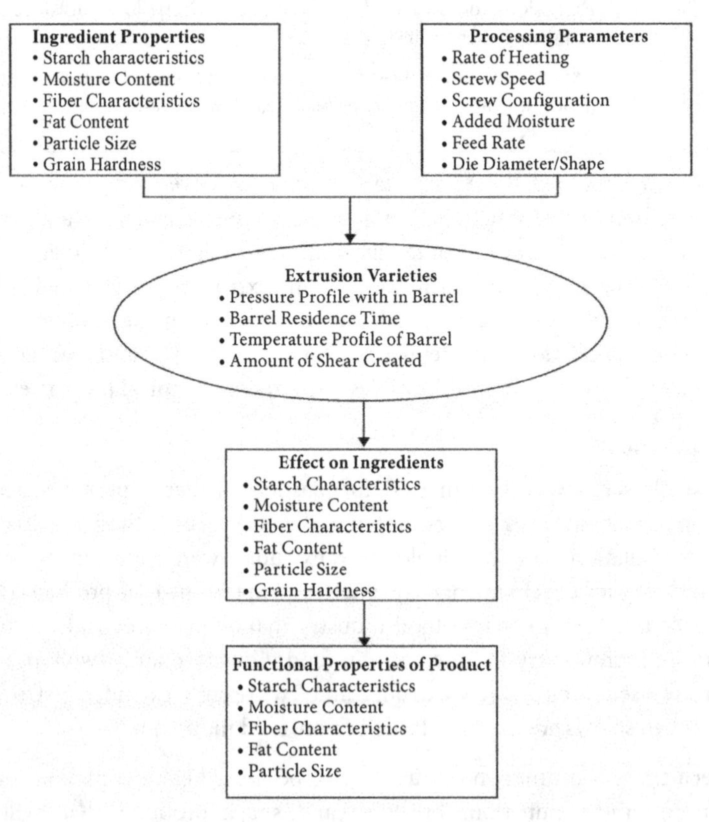

Fig 1: Potential interaction between ingredients, processing parameter and extrusion variables during extrusion cooking

As the extrusion cooking is becoming very popular researchers have transformed this process from an art to science. Extensive studies has been done and also going on presently on alteration of processing conditions, formulation of ingredients and designing of extruders for making extrusion process more efficient with enhanced applicability. Extrudates formed from fortified rice flour with jackfruit seed flour were of improved enhanced nutritional value, antioxidant and nutraceutical properties as compare to extrudates of unfortified rice flour. Processing condition including barrel temperature and screw speed considerably influence the physicochemical properties of end products (Gat and Ananthanarayan, 2015). Process variables during extrusion affect the protein digestibility as well as bioavailability of nutrients more significantly as compared with tradition cooking. It was reported that extrusion technology increases the digestibility of proteins by shearing action exhibited on protein structure and conformation during extrusion and therefore produced ready to eat products with more digestible proteins (Karkle et al, 2012). Different parameters of extrusion processing conditions affect uniquely the products properties. Navneet et al. (2010) studied effect of processing conditions and ingredients on properties of extruded products and noticed considerably higher affect on lateral expansion due to moisture content; on water absorption index due to temperature; on hardness due to speed of screw and temperature; and on sensory score due to screw speed. Extrusion cooking considerably reduced the concentration of anti-nutritional components, lowered water solubility and water absorption capacity while in vitro protein and starch digestibility increased in ready to eat products based on common beans (Karla et al, 2010).The morphological properties of fiber were also changed during extrusion cooking (Redgwell et al, 2011). Schoenfuss et al, (2013) developed expended puffed products based on modified corn-starch blended with non-fat dry milk powder twin screw extruder. A lot of work has been done on extrusion cooking regarding extrusion process, instruments and product quality. Still further research is required focussing on the interaction of ingredients that affects the end product properties; development of novel functional foods and nutraceutical products; and understanding the influence of interactions between operational parameters of extruder on nutrient retention.

Fermentation technology

Utilisation of microorganisms for development and shelf life extension of food products is the one of very old techniques of food processing. Fermentation is the biochemical reaction in which organic compounds are used metabolically to drive energy without association of any exogenous oxidizing agent. Process of fermentation includes different techniques and operations for conversion of bulky, less palatable and uneatable raw materials into palatable and value added food products with enhanced shelf life. During fermentation various inhibitory metabolites like carbon dioxide, bacteriocins, ethanol, diacetyl and organic acids including acetic acid, formic acid and lactic acid are formed which helps in preservation of fermented products. Fermented products are safer from microbial spoilage point of view in comparison to unfermented products as water activity decreased due to drying or addition of salts during fermentation process. Fermented products generally have more digestibility, improved nutritional value and better sensory properties. All these opinions have raised the interest in studying the natural fermentation processes and specifically in finding the association in multiplicity of group of fermenting microbes and their characteristics to improve the process and end product quality.

Microorganisms associated with fermented cereal foods

The groups of microorganisms generally used in production of fermented foods are bacteria, yeasts and moulds. Yeasts are being used in the production of various fermented products like wine, beer and bread since 5000 years ago. Most popular genus of yeast is *Saccharomyces* which is used for commercial production of fermented beverages and foods. In food industry, popularly used strains of *Sacchromyces* are *S. cerevisiae, S. uvarum* and *S. bayanus* and these are considered as GRAS. In bread making *S. Cerevisiae* is used dominantly while *Schizosaccharomyces pombe* and *S. boulderi* are the foremost yeasts in maize and millet based fermented beverages. The dominating bacteria used in the production of fermented foods are the *Lactobacillaceae* having the capability to form lactic acid using carbohydrates from substrates. Apart from *Lactobacillaceae*, other bacteria used in the fermentation are *Acetobacter* and *Bacillus*. *Acetobacter* are able to produce the acetic acid and used mostly in the fermentation of fruits and vegetables while *Bacillus* are able to do alkaline fermentation and generally used for fermentation of legumes. Commonly used species of *Bacillus* are *B. subtilis, B. licheniformis* and *B. pumilus*. For fermentation of protein rich foods like soybean and legumes the *B. subtilis* is the most popularly used species which is able to increase alkalinity and makes the product inappropriate for the growth of undesirable microorganisms through conversion of protein into amino acids and peptides. The fermented food products produced by *Bacillus* are pleasant organoleptically and usually consumed directly or used as flavourant and meat substitute. Some moulds are also used in fermentation of foods mainly dairy products like cheese. Microorganisms used in the production of fermented products are known as cultures. A starter culture can be defined as the group of desired microorganisms used to initiate and speed up the process of fermentation. Generally starter cultures are classified into three groups: Single strain culture (having singly strain of a species); Multi-strain cultures (composed of more than one strains from single species); and Multi-strain mixed cultures (various strains from different species).

Cereal based fermented foods

Technology of preparation of conventional foods by fermentation of cereals mainly wheat, rice, maize and sorghum is well known globally. Process of fermentation including microorganisms and substrates may vary depending on the region of the production of fermented products. Also the utilisation of fermented product is different in different parts of world such as some are used as spices, colouring and flavouring materials, beverages and breakfast or light meal foods, whereas few fermented products are consumed as staple food. Generally the process of fermentation is natural and uses mixed culture of bacteria, yeasts and fungi yet the microbiology of process is not clearly known. During the fermentation process few microorganisms may act in parallel, whereas some work in a sequential mode with varying leading flora. The type of micro-flora developed in any food during fermentation depends on various factors like temperature, moisture content, acidity, type and concentration of salts, and components of food matrix. Mixed culture is used for fermentation of rice in most of the Asian countries, whereas commercial baker's yeast or natural fermentation is used for fermentation of cereals like rye, maize, wheat and barley for development of batter for dough breads in America, Australia and Europe (Guyot, 2010). Cereals based traditional fermented foods are consumed as main meal in diet as well as complementary and infant foods in Africa. List of most commonly consumed cereal fermented foods along with their microorganisms used for fermentation and region of production is given in Table 2.

Table 2: Cereal based fermented food products

Product	Substrate	Microorganisms	Region
Anarshe	Rice	Lactic acid bacteria	India
Ang-kak	Rice	*Monascus purpureus*	Syria, Southeast Asia, China
Tape ketan	Rice or cassava	*S. cerevisiae, Hansenula anomala, Rhizopus oryzae, Chlamydomucor oryzae, Mucor, Endomycopsis fibuliger*	Indonesia
Miso	Rice with soy beans or barley	*A. oryzae, Torulopsis etchellsii, Lactobacillus*	Japan, China
Lao-chao	Rice	*Rhizopus oryzae, R. chinensis, Chlamydomucor oryzae, Saccharomycopsis*	China, Indonesia
Bhattejaanr	Rice	*Hansenula anomala, Mucor rouxianus*	India, Sikkim
Kichudok	Rice	*Saccharomyces*	Korea
Dhokla	Rice or wheat and bengal gram	*L. mesenteroides, Streptococcus faecalis Torulopsis candida, T. pullulans*	Northern India
Khanom-jeen	Rice	*Lactobacillus, Streptococcus*	Thailand
Dosa	Rice and bengal gram	*L. mesenteroides, Streptococcus faecalis, Torulopsis candida, T. pullulans*	India

Product	Substrate	Microorganisms	Region
Puto	Rice, sugar	*L. mesenteroides, Strepromyces faecalis, yeasts*	Philippines
Idli	Rice and black gram	*L. mesenteroides, Streptococcus faecalis, Torulopsis, Candida, Tricholsporon pullulans*	South India, Sri Lanka
Sierra rice	Rough rice	*Aspergillus flavus, A. candidus, Bacillus subtilis*	Ecuador
Koko	Maize	*Enterobacter clocae, Acinetobacter., Lactobacillus platarum, L. brevis, S. cerevisiae, Candida mycoderma*	Ghana
Chicha	Maize	*Penicillium, Aspergillus, yeasts, bacteria*	Peru
Pozol	Maize	*Lactococcus lactis, Streptococcus suis, Lactobacillus plantarum, Lactobacillus casei, Lactobacillus alimentarium, Lactobacillus delbruekii and Clostridium* sp.	South-eastern Mexico
Uji	Maize, Sorghum, millet	*L. mesenteriodes, Lactobacillus platarum*	Kenia, Uganda, Tanganyika
Atole	Maize	*Lactic acid bacteria*	Southern Mexico
Jamin-bang	Maize	*Yeasts, bacteria*	Brazil
Chee-fan	Soybean, wheat	*Mucor, Aspergillus glaucus*	China
Nan, Kulcha	Unbleached wheat flour	*S. cerevisiae,* lactic acid bacteria	India, Pakistan, Iran, Afghanistan,
Hamanatto	Wheat, soybeans	*Aspergillus oryzae, Streptococcus, Pediococcus*	Japan
Jalebies	Wheat flour	*S. bayanus*	India, Pakistan, Nepal,
Mantou	Wheat flour	*Saccharomyces*	China
Kishk	Wheat and milk	*Lactobacillus plantarum, L. brevis, L. casei, Bacillus subtilis and yeasts*	Egypt, Syria, Arabian countries
Injera	Sorghum, maize or wheat	*Candida guilliermondii*	Ethiopia
Nasha	Sorghum	*Streptococcus, Lactobacillus, Candida, S. cerevisiae*	Sudan
Rabdi	Maize and buttermilk	*Penicillium acidilactici, Bacillus, Micrococcus*	India

Blandino *et al.*, 2003; Das *et al.*, 2012; Soni & Sandhu, 1990

Recent advances

In earliest times flavour development and shelf life extension were the main objectives of food fermentation process. Industrialization of native fermented products has increased the chances for exploring the involvement of biotechnologies in improvement of production methods and food quality. Biotechnology has played key role in formulation, shelf life extension, nutritional enrichment and value addition of fermented foods. Major outcomes of application of these technologies are the assurance of food safety regarding physical, microbial and chemical hazards and quality of end food products. Exploring the knowledge of microbiology in food processing regarding recognition of novel fermenting species was a boon to food industry for improving the food quality in a number of ways. Since very long, biotechnology has been a supportive base in development of new functional foods, organoleptic properties improvement, bio-preservation and enzyme modification. New techniques like genetic engineering in the field of food biotechnology will direct the area of functional foods.

Backslopping is technique of using a small quantity of previously fermented ready product as inoculums for the next batch. This method has been used since the beginning of production of fermented products in the history and still in use but role of microorganism in culture is not clearly known. In modern food fermentation industry, functional starter cultures have been used very commonly since last two to three decades. Functional starter culture is the culture containing at least on inherent functional property with it. Probiotic strains come under the category of functional starter cultures where careful selection of starter is done to produce end product with desirable health properties.

The functionality of starter culture depends on strain's growth, substrate and fermentation conditions. Therefore, the biochemical and functional characteristics of yeasts used in preparation of fermented foods been intensively studied to develop desired starter culture (Montet and Ray, 2016). Genetic engineering is considered safe and has been implemented in the development of starter culture to be used in food fermentation. One example of genetic engineering for food safety was the removal of the D-lactate dehydrogenase (ldhD) gene from *Lactobacillus johnsonii La1*, which demonstrated the exclusion of the undesired D-isomer of lactate and leaving only the desired L-lactate (Mollet, 1999). The conversion of sugars to acids and metabolites with unique flavour through pyruvate is the example of metabolic engineering of microorganisms to alter the properties of starter culture. Bacterial genomics also play important role in improving the culture quality. Study of collected biological data and achieving the complete genome sequence for population of microorganisms including LAB provided a direction in physiological studies, selection of mutant, and selection and implementation of genetic and protein engineering techniques for starter culture development. The genome sequence of *Lactococcus lactis* exposed a number of unpredicted genetic and metabolic potentials (Bolotin et al, 2001). Safety aspect of microorganisms used in food fermentation is an important issue as it directly relates to consumer health and quality of product as well. Commonly used microorganisms used in food industry for fermentation come under the category of GRAS. However, it is not necessary that any microorganism is GRAS for one food product must be same for all or any other food products. European Food Safety Authority (EFSA) started an assessment Qualified Presumptions of Safety (QPS) for analysis of microorganisms needing a market approval in November 2007 (Vogel et al, 2011). The QPS assessment was done on criteria including taxonomic level, record of previous use and recognition of potential safety fears. A list of species wise microorganisms was organized which are considered safe for use and this list is being updated annually. To consider any microorganism safe for human use the important factors to be considered are its undesirable characteristics, toxicity of produced metabolites, chances of infection, and resistance to antibiotic.

Lactic acid bacteria are used globally for fermentation of wide range of food products due to their ability for biosynthesis of a variety of compounds as metabolic end products or secondary metabolites. LAB also have potential to enhance the flavour, nutritional value and textural properties of products. These bacteria are also able to produce considerably high quantity of proteins or enzymes. LAB accelerate the conversion process of lactose into sugar alcohols which is required for the production of low calorie sweeteners. The functional bio-molecules produced by lactic acid bacteria play important role in quality enhancement of cereal based fermented beverages (Waters et al, 2015). Therefore, extensive use of LAB in the food industry for production of healthier and functional foods makes this technology of using these microorganisms as cell factories of great importance. Immobilization and encapsulation of cell are the new technologies of interest in fermented food industry. Encapsulation technique uses cells enrobed within a gel-matrix, where fermentation is carried out by the metabolic activity of encapsulated cells. The encapsulation of bacteria is an enhanced technology in food science which mainly focused on cost-effectiveness, rapid, non destructive and food grade purity that will improve the quality of end product. The technology of micro-encapsulation is the further advancement of immobilization technology and recently used in fermentation processing for microencapsulation of live probiotic bacterial cells (Mortazavian et al, 2016).

Technologies for enhancement of bioactive components

Food consumed is one of the most important factors for growth of human body and maintenance of good health condition. Therefore the interest of nutritionists, food scientists and technologists, consumers, and governments is increasing in investigating the correlation between health condition and food consumed. Cereal grains contain various nutrients, antioxidants, dietary fibres and phytochemicals which are helpful in prevention of diseases and support good health. Studies showed that consumption of whole grains is more beneficial to health than processed grains and these health benefits are credited to the collective effects of nutrients, dietary fibres and bioactive components positioned mainly in germ portion and outer bran layer. The reason behind it is the removal of bran and germ part containing bioactive compounds during milling or polishing for production of refined flours. Loss of bioactive compounds also takes place during various processing treatments like baking, cooking, extrusion, and puffing generally followed for development of cereal based products. Thus, food scientists and nutritionists are working to find out some new approaches and techniques for the improvement of level of these bioactive compounds, dietary fibre contents and nutritional components in processed cereal foods. Some strategies and processing treatments have been practiced to deal with insufficiencies regarding contents of nutritional components, bioactive compounds and dietary fibres of refined flour and to enhance the quality of end products based on cereals (Figure 2). Supplementation of cereal based products with ingredients of fruit and vegetables, seeds and legumes, and plants parts or extracts is commonly used technology for enrichment of cereal products. Utilisation of composite flours and whole grain flours is another strategy to deal with deficiencies of useful compounds in cereal products. Some processing treatments like fermentation, sprouting and enzymatic treatments are also proposed in literature to improve the bioavailability of bioactive compounds and micronutrients of cereal foods (Singh et al, 2016). Innovative breeding methods, bio-fortification of crops and genetic engineering are the new technologies for production of cereal crops rich in bioactive compounds.

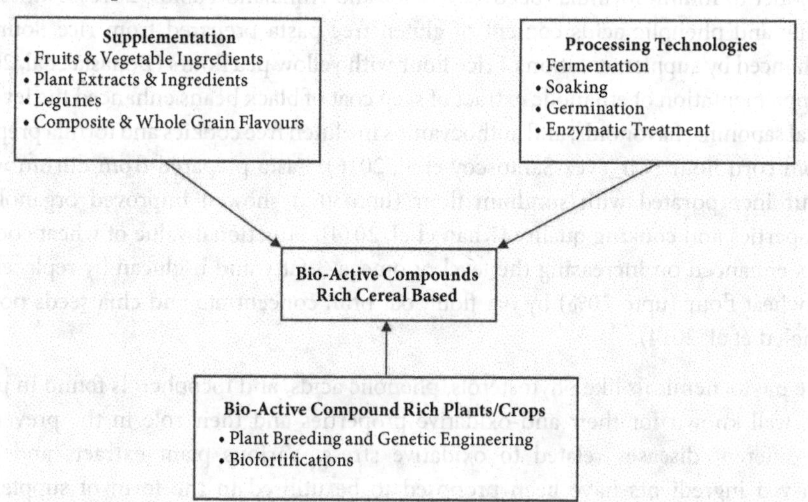

Fig. 2: Various technologies for development of cereal based products rich in bioactive compounds

Supplementation of cereal based products

Fruits and vegetables and their by products like peels, seeds and pulp residues are rich in bioactive compounds and have been successfully used for supplementation purpose in development of cereal-based foods. Usually, fruits and vegetables and their by-products are dried and grinded to powder form to make them suitable ingredients for supplementation. The enrichment is chiefly credited to higher contents of bioactive compounds and dietary fibre in supplemental ingredients from fruits and vegetables as compared to processed cereal grain and refined flour. Legumes are good source of nutrients and non-nutritive bioactive compounds as well as contain higher level of resistant starch than cereal grains, therefore can be used for supplementation of cereal products. The intensity of dietary fibres and bioactive compounds enrichment in the final products depends on type of processing for food preparation as well as level and type of ingredients used for supplementation. Level of bioactive compounds like phenolic compounds, terpeniods and catotenoids was enhanced in extrudates from corn flour by supplementation with 19 to 21% pumpkin powder (Hong et al, 2015). Mango pulp fibre is a by product in mango processing after juice extraction and can be used for supplementation of foods as it is rich source of dietary fibres and other phenolic compounds. The supplementation of wheat flour for muffin preparation by incorporation of dried powder of mango pulp fibre resulted in increased level of dietary fibre, total phenolics, b-carotene, and lutein in muffins (Sudha et al, 2015). Lipid oxidation of muffine was inhibited by addition of extract of pineapple by-product to muffin formula (Segovia Gómez and Almajano Pablos, 2016). Nutritional value and phenolic acids content of gluten free pasta prepared from rice flour was enhanced by supplementation of rice flour with yellow pea flours (Bouasla et al, 2016). Supplementation of ethanolic extract of seed coat of black beans enhanced the levels of total saponins, flavonoids, and anthocyanins in gluten free cookies and tortilla prepared from corn flour (Ch_avez-Santoscoy et al, 2016). Pasta prepared from durum wheat flour incorporated with sorghum flour (upto30%) showed improved organoleptic properties and cooking quality (Khan et al, 2014). Functional value of wheat cooking was enhanced on increasing the level of omega-3 fatty and b-glucan by replacement of wheat flour (upto 20%) by oat flour, oat bran concentrate and chia seeds powder (Inglett et al, 2014).

The phytochemicals like phytosterols, phenolic acids, and tocopherols found in plants are well known for their anti-oxidative properties and their role in the prevention of different diseases related to oxidative stress. Various plant extracts and plant derived ingredients have been proposed to be utilised in the form of supplement for enhancement of dietary fibre and bioactive compounds levels in food products based on cereals. Green tea extracts dried by spray drier and freeze drier added in wheat flour for biscuits preparation enhanced the catechin level in dough without influencing the quality of end product. However, the catechin level reduced on increasing the temperature during baking (Sharma and Zhou, 2011). In another study, green tea extracts in encapsulated and unencapsulted forms added in dough for bread development increased the total phenolics level in end product with no effect on volume and crumb firmness (Pasrija et al, 2015). Nutritional value of buns prepared from wheat flour supplemented with leaves powder of curry and coriander improved as compared to wheat flour buns (Sudha et al, 2014).

Processing technologies

Quantity and quality of nutrients and other non-nutritive components in end products depends on the initial raw material as well as processing conditions of food preparation. Different handling and processing treatments influence quality of end products differently which may be positive or negative influence. There are some processing methods which are reported to be effective in improving the nutritional value and bioactivity of end food products. Therefore, levels of bioactive compounds, dietary fibres and nutrients can be improved in cereal based products by selecting appropriate processing methods and optimised processing conditions.

The processes of germination, fermentation and soaking of cereal grains are reported to increase the content of nutrients and bioactivity in end products. Wheat flour showed increased levels of nutrients, dietary fibres, total phenolic compounds and antioxidant activity after germination of wheat grains (Van Hung et al, 2015). Enhanced antioxidant activity was observed in barley after germination which could be due to the enzymatic activities during germination causing degradation of polysaccharides and increased hydrogen bonding due to exposure of hydroxyl groups (Ahmad et al, 2016). Generally, the increment in levels of bioactive compounds during germination is associated with enzymatic activities causing freeing of bound phenolics and glycosylation reactions.

In another study, improvement in concentration of nutrients and increased bioactivity was observed in various cereals including wheat, maize, millet, oat, sorghum and rice after solid-state fermentation (Zhai et al, 2015). Study conducted on preparation of fermented wheat products using optimal combination of flour-microbial strains showed improved functional and organoleptic properties of end products. It was reported that enhancement in bioactive compounds and betterment in sensory properties depends on wheat variety and microbial strain used for fermentation (Ferri et al, 2016). Air classification technology produced coarse fractions with higher antioxidant activities, total phenolics acids and beta-glucans than whole meal of barely (G_omez-Caravaca et al, 2015). Similarly, stone milling method produced flour with higher nutrients and bioactive compounds than roller milling method in case of durum wheat (Ficco et al, 2016).

Conclusion

Cereals and cereal based products contain an important place in human diet and have been consumed in different ways in different regions of the world since ancient times. Urbanisation and modernisation of society has changed the food pattern of human and now trend is increasing towards ready to eat or ready to cook, convenient, value added foods with lesser chemical additives and enhanced sensory characteristics. This tendency has emphasised on finding the connections of ingredients and processing technologies used in food preparation. Development of new technologies such as extrusion cooking, bio-fortification of crops for enhancement o bioactive compounds and genetic engineering in plants and microorganisms used for food preparations are the witnesses of evolution in cereal products industry. Modification in traditional processing methods and food products being done on the basis of knowledge, requirements and tradition in different parts of world is also the case of evolution in the area covering cereal products. For today's food scientists and technologists, having the advance knowledge of cereal-food technologies, it is crucial to constantly improve the efficiency of existing practices, development of novel competitive cereal based food products and to envisage future trends in the industry.

References

Abu-ghoush, M., Herald, T.J., and Aramouni, F.M. (2010). Comparative study of egg white protein and egg alternatives used in an angel food cake system. *Journal of Food Processing and Preservation,* 34, 411-425.

Ahmad, M., Gani, A., Shah, A., Gani, A., and Masoodi, F. (2016). Germination and microwave processing of barley (*hordeum vulgare* l) changes the structural and physicochemical properties of β-d-glucan and enhances its antioxidant potential. *Carbohydrate polymers, 153*, 696-702.

Asghar, A., Anjum, F.M., Allen, J.C., Daubert, C.R., and Rasool, G. (2009). Effect of modified whey protein concentrates on empirical and fundamental dynamic mechanical properties of frozen dough. *Food Hydrocolloids,* 23(7), 1687-1692.

Bilgin, B., Daglioglu, O., and Konyali, M. (2006). Functionality of bread made with pasteurized whey and/or buttermilk. *Italian Journal of Food Science, 18*(3).

Blandino, A., Al-Aseeri, M., Pandiella, S., Cantero, D., and Webb, C. (2003). Cereal-based fermented foods and beverages. *Food Research International,* 36(6), 527-543.

Bolotin, A., Wincker, P., Mauger, S., Jaillon, O., Malarme, K., Weissenbach, J., Sorokin, A. (2001). The complete genome sequence of the lactic acid bacterium lactococcus lactis ssp. Lactis il1403. *Genome Research, 11*(5), 731-753.

Bouasla, A., Wójtowicz, A., Zidoune, M.N., Olech, M., Nowak, R., Mitrus, M., and Oniszczuk, A. (2016). Gluten-free precooked rice-yellow pea pasta: Effect of extrusion-cooking conditions on phenolic acids composition, selected properties and microstructure. *Journal of Food Science,* 81(5), C1070-C1079.

Ch_avez-Santoscoy, R. A., J. A. Guti_errez-Uribe, S. O. Serna-Saldivar, and E. Perez-Carrillo. 2016. Production of maize tortillas and cookies from nixtamalized flour enriched with anthocyanins, flavonoids and saponins extracted from black bean (*Phaseolus vulgaris*) seed coats. *Food Chem.* 192, 90–7.

Das, A., Raychaudhuri, U., and Chakraborty, R. (2012). Cereal based functional food of indian subcontinent: A review. *Journal of Food Science and Technology,* 49(6), 665-672.

Divya, N., and Rao, K.J. (2010). Studies on utilization of indian cottage cheese whey in wheat bread manufacture. *Journal of Food Processing and Preservation,* 34(6), 975-992.

Ferri, M., Serrazanetti, D.I., Tassoni, A., Baldissarri, M., and Gianotti, A. (2016). Improving the functional and sensorial profile of cereal-based fermented foods by selecting lactobacillus plantarum strains via a metabolomics approach. *Food research international,* 89, 1095-1105.

Ficco, D.B.M., De Simone, V., De Leonardis, A.M., Giovanniello, V., Del Nobile, M.A., Padalino, L., De Vita, P. (2016). Use of purple durum wheat to produce naturally functional fresh and dry pasta. *Food Chemistry,* 205, 187-195.

G_omez-Caravaca, A. M., V. Verardo, T. Candigliota, E. Marconi, A. Segura-Carretero, A. Fernandez-Gutierrez, and M. F. Caboni. 2015. Use of air classification technology as green process to produce functional barley flours naturally enriched of alkylresorcinols, b-glucans and phenolic compounds. *Food Res. Int.* 73,88–96.

Gat, Y., and Ananthanarayan, L. (2015). Effect of extrusion process parameters and pregelatinized rice flour on physicochemical properties of ready-to-eat expanded snacks. *Journal of Food Science and Technology,* 52(5), 2634-2645.

Guyot, J.-P. (2010). Fermented cereal products. *Fermented Foods and Beverages of the World. London:* CRC Press (Taylor and Francis Group), 247-261.

Hong, F.L., Peng, J., Lui, W.B., and Chiu, H.W. (2015). Investigation on the physicochemical properties of pumpkin flour (*Cucurbita Moschata*) blend with corn by single-screw extruder. *Journal of Food Processing and Preservation,* 39(6), 1342-1354.

Huschka, B., Challacombe, C., Marangoni, A.G., and Seetharaman, K. (2011). Comparison of oil, shortening, and a structured shortening on wheat dough rheology and starch pasting properties. *Cereal Chemistry,* 88(3), 253-259.

Inglett, G.E., Chen, D., Liu, S.X., and Lee, S. (2014). Pasting and rheological properties of oat products dry-blended with ground chia seeds. *LWT-Food Science and Technology,* 55(1), 148-156.

Karkle, E. L., Keller, L., Dogan H. and Alavi, S. (2012). Matrix transformation in fiber-added extruded products: Impact of different hydration regimens on texture, microstructure and digestibility. *J. Food Eng. 108,* 171–182.

Karla, A., Batista, S. H., and Prud^encio, K. F. F. (2010). Changes in the functional properties and antinutritional factors of extruded hard-to-cook common beans (*Phaseolus Vulgaris* L.). *J. Food Sci. 75(3),* 286–290.

Khan, I., Yousif, A.M., Johnson, S.K., and Gamlath, S. (2014). Effect of sorghum flour addition on in vitro starch digestibility, cooking quality, and consumer acceptability of durum wheat pasta. *Journal of Food Science,* 79(8), S1560-S1567.

Manley, D. (2011). *Manley's Technology of Biscuits, Crackers and Cookies*: Elsevier.

Martha, G. Ruiz-Gutiérrez, Miguel Á. Sánchez-Madrigal, and Armando Quintero-Ramos. (2017). The Extrusion Cooking Process for the Development of Functional Foods, Extrusion of Metals, Polymers, and Food Products. Sayyad Zahid Qamar, Intech Open.

Mazo, V.K., Gmoshinski, I.V., and Zorin, S.N. (2007). New food sources of essential trace elements produced by biotechnology facilities. *Biotechnology Journal: Healthcare Nutrition Technology,* 2(10), 1297-1305.

Mollet, B. (1999). Genetically improved starter strains: Opportunities for the dairy industry. *International Dairy Journal,* 9(1), 11-15.

Montet, D., and Ray, R.C. (2016). *Fermented Foods, Part I: Biochemistry and Biotechnology*: CRC Press.

Mortazavian, A., Moslemi, M., and Sohrabvandi, S. (2016). Microencapsulation of probiotics and applications in food fermentation. *Fermented Foods. Part, 1,* 185-210.

Moscicki, L., and van Zuilichem, D.J. (2011). Extrusion-cooking and related technique. *Extrusion-cooking techniques: applications, theory and sustainability. Wiley, Weinheim,* 1-24.

Navneet, K., Sarkar, B. C., and Sharma, H. K. (2010). Development and characterization of extruded product of carrot pomace, rice flour and pulse powder. *African J. Food Sci. 4(11),* 703–717.

Nunes, M.H.B., Moore, M.M., Ryan, L.A., and Arendt, E.K. (2009). Impact of emulsifiers on the quality and rheological properties of gluten-free breads and batters. *European Food Research and Technology,* 228(4), 633-642.

Pasrija, D., Ezhilarasi, P., Indrani, D., and Anandharamakrishnan, C. (2015). Microencapsulation of green tea polyphenols and its effect on incorporated bread quality. *LWT-Food Science and Technology,* 64(1), 289-296.

Patton, J., and Nabors, L.O.B. (2011). Polyols: Sweet oral benefits. *Journal of International Oral Health,* 3(5), 1.

Redgwell, R., Curti, D., Robin, F., Donato, L., and Pineau, N. (2011). Extrusion-induced changes to the chemical profile and viscosity generating properties of citrus fiber. *Journal of Agricultural and Food Chemistry,* 59(15), 8272-8279.

Robert, H., Gabriel, V., and Fontagné-Faucher, C. (2009). Biodiversity of lactic acid bacteria in french wheat sourdough as determined by molecular characterization using species-specific pcr. *International Bournal of Food Microbiology,* 135(1), 53-59.

Rodríguez-García, J., Puig, A., Salvador, A., and Hernando, I. (2012). Optimization of a sponge cake formulation with inulin as fat replacer: Structure, physicochemical, and sensory properties. *Journal of Food Science,* 77(2), C189-C197.

Schoenfuss, T., Tremaine, A., Evenson, K., and Maher, M. (2013). Twin-screw extrusion puffing of non-fat dry milk powder. Pp: 1-30, Agriculture Utilization Research Institute, (https:// www. auri.org/ assets/2013/ 11/2009103. Schoenfuss.pdf).

Segovia Gómez, F., and Almajano Pablos, M.P. (2016). Pineapple waste extract for preventing oxidation in model food systems. *Journal of Food Science, 81*(7), C1622-C1628.

Sharma, A., and Zhou, W. (2011). A stability study of green tea catechins during the biscuit making process. *Food Chemistry,* 126(2), 568-573.

Singh, A., Sharma, V., Banerjee, R., Sharma, S., and Kuila, A. (2016). Perspectives of cell-wall degrading enzymes in cereal polishing. *Food Bioscience,* 15, 81-86.

Soni, S., and Sandhu, D. (1990). Indian fermented foods: Microbiological and biochemical aspects. *Indian Journal of Microbiology,* 30(2), 135-157.

Stortz, T.A., Zetzl, A.K., Barbut, S., Cattaruzza, A., and Marangoni, A.G. (2012). Edible oleogels in food products to help maximize health benefits and improve nutritional profiles. *Lipid Technology, 24*(7), 151-154.

Sudha, M. L., Indumathi, K., Sumanth, M. S., Rajarathnam, S., and Shashirekha, M. N. (2015). Mango pulp fibre waste: characterization and utilization as a bakery product ingredient. *J. Food Meas. Charact.* 9, 382–388.

Sudha, M., Rajeswari, G., and Rao, G.V. (2014). Chemical composition, rheological, quality characteristics and storage stability of buns enriched with coriander and curry leaves. *Journal of Food Science and Technology, 51*(12), 3785-3793.

Sudha, M., Vetrimani, R., and Leelavathi, K. (2007). Influence of fibre from different cereals on the rheological characteristics of wheat flour dough and on biscuit quality. *Food Chemistry,* 100(4), 1365-1370.

Van Hung, P., Maeda, T., and Morita, N. (2015). Improvement of nutritional composition and antioxidant capacity of high-amylose wheat during germination. *Journal of Food Science and Technology,* 52(10), 6756-6762.

Van Riemsdijk, L.E., Van der Goot, A.J., and Hamer, R.J. (2011). The use of whey protein particles in gluten-free bread production, the effect of particle stability. *Food Hydrocolloids,* 25(7), 1744-1750.

Vogel, R.F., Hammes, W.P., Habermeyer, M., Engel, K.H., Knorr, D., and Eisenbrand, G. (2011). Microbial food cultures–opinion of the senate commission on food safety (sklm) of the german research foundation (dfg). *Molecular Nutrition and Food Research,* 55(4), 654-662.

Waters, D.M., Mauch, A., Coffey, A., Arendt, E.K., and Zannini, E. (2015). Lactic acid bacteria as a cell factory for the delivery of functional biomolecules and ingredients in cereal-based beverages: A review. *Critical Reviews in Food Science and Nutrition,* 55(4), 503-520.

Zhai, F.-H., Wang, Q., and Han, J.-R. (2015). Nutritional components and antioxidant properties of seven kinds of cereals fermented by the basidiomycete agaricus blazei. *Journal of Cereal Science,* 65, 202-208.

Zoulias, E., Oreopoulou, V., and Tzia, C. (2002). Textural properties of low-fat cookies containing carbohydrate-or protein-based fat replacers. *Journal of Food Engineering,* 55(4), 337-342.

7

Technological Evaluation of Milling Operations

Arvind, Shikha Pandhi and Veena Paul*

Department of Dairy Science and Food Technology, Institute of Agricultural Sciences, Banaras Hindu University, Varanasi, Uttar Pradesh

Introduction

The farm to fork approach utilizes agricultural products as raw material for processing by industries to obtained value-added products (Mijinyawa et al., 2007). Processing that transforms animal, vegetable, or marine materials into value-added food products for consumers and forms an integral part of agriculture for storage or size reduction into fine particles. One of the major problems associated with processing is the unavailability of processing facilities including appropriate engineering materials for the transformation of agro-based produce into different forms (Ogwuagwu, 2007). This transformation process mainly includes operations like particle size reduction, milling or comminution, etc. utilizing different size reduction forces without making alterations in chemical properties (Enrique, 2012). This enhances the eating quality or appropriateness of foods for advance processing with a more varied range of available products.

Grains are dried small seeds with adhered or removed husks or outer layers processed for consumption (FAO, 2011). The two major types of commercial grain crops include cereals (such as millet, corn, sorghum, wheat, rye and so on) and legumes (such as beans, groundnuts and soya beans). These grain crops are subjected to a milling operation to break them into smaller pieces or more practically appropriate form. The grinding process involves the application of different mechanical forces to overcome the interior bonding forces bringing change in the state of solid grains to different desirable forms such as flour or meal (Kaul and Egbo, 1985). As a result, the physical state of these grains is modified due to a reduction in size and change of shape (Ryan and Spencer, 2008). Most of the actual milling processing that is used in various industries is based on simple operations. But now various technological progress has been initiated that involves the adoption of computer control for various milling operations to improve efficiency and quality while minimizing labor requirements. Hence, the present text highlights the various technological advancements and evaluation techniques employed in milling operations.

Technological evaluation of the performance of milling machine is important to determine its efficiency and suitability for milling of grains involves applications of a series of principles, techniques, tools, and, methods which efficiently assess the significant importance of the technology and its involvement in enhancing the efficiency of milling operations. It involves the selection of innovative ideas, evaluating innovative as well as not-innovative products and technologies from technical, market and consumer perspectives and reconciles the results within a valid methodology. This chapter speculates various milling operations of grains that are carried out in the milling industry. Also, it highlights the various technological advancements that have been made along with the measures to evaluate efficiency.

Milling of grains

Milling of grains is defined as a process of breaking down of coarse grains into fine particles. In other words, milling can be defined as a transformation operation that converts the raw material into a primary product for its further processing. Foodgrains such as cereals (wheat, rice, wheat, maize, sorghum, and millets), pulses (chickpea, urad bean, pigeon pea, and mung bean), and oilseeds (soybean, groundnut, sesame, rapeseed, and mustard) could be transformed using milling operations for further processing. Foodgrains play an important role in fulfilling the food, feed, and seed requirements of the people. Milling of these food grains leads to their conversion in suitable products. Milling operation is efficient only if processing technology and equipment used are according to crop and their intended end-use. This may include primary (milling of paddy into rice) as well as secondary processing of grains (milling of chickpea, splits into Besan). The main objective of milling of grains is to separate the grains from stems because, after the removal of outer layered husk, grains are ready to cook. Cereal grains such as wheat, millets, and sorghum are milled without husk whereas for the milling of maize the outer covering sheath of the maize cob should be removed. Cereal grains like rice, barley, and oats with husk tightly attached to the grains. So, firstly the husk should be removed by rubbing the grains in the rubber roller to get high out-turn of the whole grains. After removal of the husk, these cereals grains are milled in a similar way as wheat, millets, and sorghum grains. Rather than a husk, the other parts of cereal grains that must be removed are germ and the outer layer, collectively known as 'Bran'. This bran is high in lipids which tend to oxidize to produce off-flavors and are less digestible. So, the purpose of milling is to remove the husk as well as bran. As the removal of the bran, during the milling process produces high-quality flour (Dobraszczyk and Dendy, 2001).

Wheat milling

Wheat ranks second important food grains of India followed by rice. Approximately 10% of the post-harvest loss is there during harvesting, storage, and transportation. There are different varieties of wheat required for the production of various wheat-based products. Wheat is milled with a moisture percentage of 14%. The structure of wheat grain is an important part of the milling of wheat. The grain of wheat consists of outer covering the pericarps and testa which is hard and indigestible, an aleurone layer, which contains a higher proportion of protein than flour, an embryo attacked to a small structure, the scutellum at the lower end of the grain, and finally the endosperm comprising of 85 percent of the whole grain from which the flour is derived. It is important to mention that the milling of wheat is a physical process. The wheat bran is composed of 12% bran, 3% germ, and 85% endosperm.

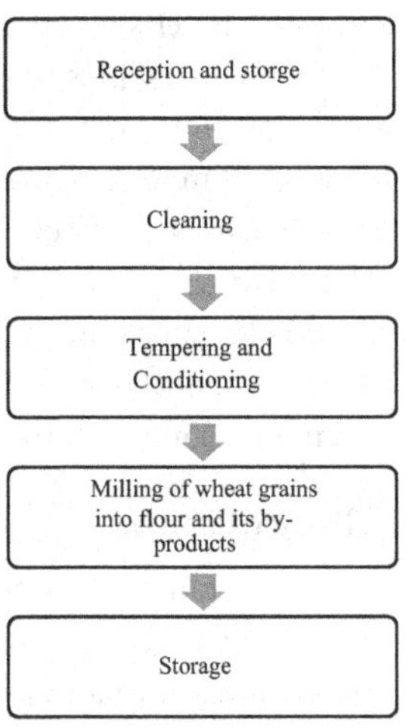

Fig. 1: Process flow chart of wheat milling

Milling of wheat aims to reduce the size of grain by separating the endosperm from the bran layer and germ layer. Generally, wheat mills yield 70% of wheat flour. The process flow chart of wheat milling is shown in Figure 1. The wheat to be milled should be free from contamination and the protein content must be in range depending upon the products to be prepared. The protein content of the flour is always lower as compared to wheat grain. Thus, the milling of wheat needs timely inspection and grading.

Hence, this grading system leads to the selection of sound grains that are free from contamination and store with a proper moisture content of 14%. During the storage of grains, the impurities are removed by the removal of an adhered layer of dust on the surface of grains. Stones, small particles and sand are removed by passing air current to the wheat grains due to which light particles are removed. Then the wheat grains are passed through a magnetic separator for the removal of metal impurities. The next step is tempering, defined as, the addition of water to the endosperm and bran to raise the moisture level. Tempering results in leathery bran along with less vitreous endosperm. Tempering is an important aspect that enhances milling efficiency. Conditioning is done to provide heat which is required after quick diffusion of water into bran and endosperm. There is four conditioning section in wheat miller. In the first section, heat (45°C temperature) is provided which affects the quality of gluten. The other section adds moisture and holds the wheat for the proper temperature. The next section provides cooling for one and half hours or less to the wheat grains at room temperature and the last section contains a holding bin where the wheat grains are held for 8-18 hours for equilibrating the moisture before milling.

The milling of wheat flour consists of the three main steps as shown in Figure 2. The first step is the purification system which is not favored by millers and may absent in various milling processes. The second step is the breaking system, where the grains are broke down and separation of endosperm occurs by the help of roller miller with saw tooth profile surface. When the grains come in contact with the break system, it splits the grains and releases a significant amount of material by the process of sieving into the purification and reduction section. This produces wheat flour in a small amount and the process is repeated for 4 – 5 times. When the expected quantity of material obtained from split grains is released, the residual material is then discharged in the form of a co-product as wheat feed stream. Then it further moves towards the reduction section, which is the chief flour producing section. This section also manipulates the desirable properties of flour such as mechanical starch damage (Owens, 2001). The reduction section comprises of a series of roller mills and sifters. Materials from break and purification system are then subjected towards these reduction rollers for size reduction, coupled with sieving equipment to remove residual impurities. In the reduction system, the roller mills used are different than those used in break system. As the roller mills used in the reduction system consists of smooth-surfaced rollers moving with lower differential speeds. Materials that do not attain the size are further reduced by grinding and sieving process has done repeatedly approximately 11 times till it achieved fine particle size flour.

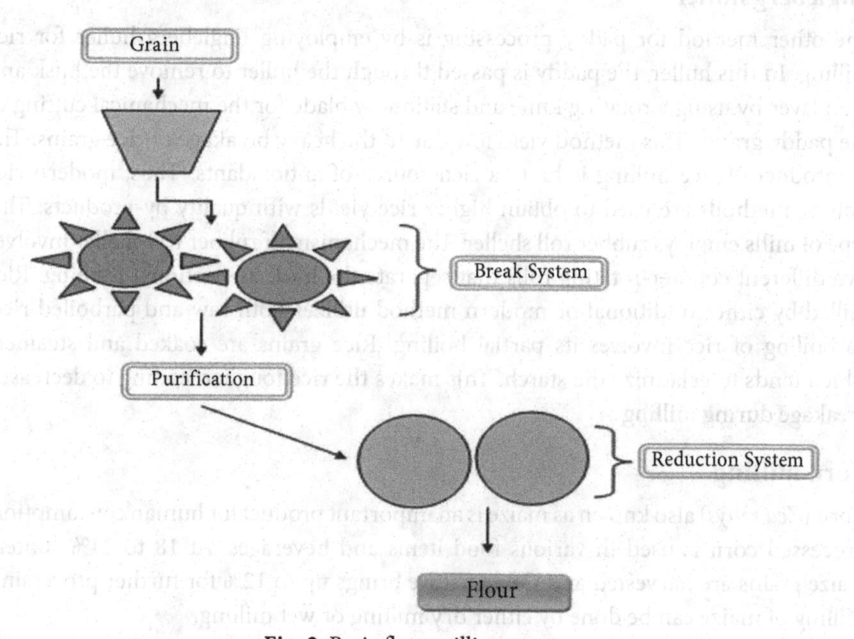

Fig. 2: Basic flour milling process

Rice milling

Paddy (*Oryza sativa L.*) is one of the most important grains in the world and almost 2/3rd of the population worldwide adopted it as their staple food. India ranks second in rice production. There are several layers in paddy grains viz., husk, bran, starchy endosperm, and the germ layer. The husk is the coating for the seeds also known as hull. The outer layer of paddy grain, bran, is comprised of aleurone, nucleus, seed coat, and pericarp. The germ is an internal layer of whole-grain, obtained as a by-product of paddy milling. In the process of paddy milling, rice grain is transformed to improve the recovery without breakage of the kernel. During the rice milling process, the removal of the husk layer results in brown rice. Brown rice is further processed for the removal of the bran layer which results in white rice.

Methods of rice milling

Hand pound method

Milling of rice using hand pound is a traditional method, which involves pounding of paddy in stone or pestle and mortar. Then the paddy grains are winnowed, which leads to the separation of de-husked and unhusked kernels. Further, unhusked grains are pounded. This method is burdensome, takes a longer time, and labor-intensive.

Engleberg huller

The other method for paddy processing is by employing Engleberg huller for rice milling. In this huller, the paddy is passed through the huller to remove the husk and bran layer by using a rotating knife and stationary blade for the mechanical cutting of the paddy grains. This method yield low due to the heavy breakage of rice grains. The by-product of rice milling is bran, a rich source of antioxidants. Thus, modern rice milling methods are used to obtain higher rice yields with quality by-products. This type of mills employs rubber roll sheller. The mechanism of rubber roll sheller involves two different counter-rotating rolls that separate the husk by frictional rubbing. Rice milled by either traditional or modern method utilizes both raw and parboiled rice. Parboiling of rice involves its partial boiling. Rice grains are soaked and steamed, which tends to gelatinize the starch. This makes the rice tougher leading to decreased breakage during milling

Corn milling

Corn (*Zea mays*) also known as maize is an important product for human consumption. Processed corn is used in various food items and beverages. At 18 to 24% content maize grains are harvested and the moisture brings up to 12% for further processing. Milling of maize can be done by either dry milling or wet milling.

Dry milling

The dry milling of maize can be done by either the traditional method or the modern method. In the traditional method, the maize kernel is firstly grounded into a meal by employing a stone grinder. In this process, the germ layer is not removed and is rich in fiber and protein. Then, the germ layer is removed. In the modern method, the maize kernel is firstly moistened with water and then tempered for equilibrium. Tempering is done to remove all the germ layer so that a maximum amount of endosperm can be obtained without any dark specks. Then the moistened kernels are dried followed by removal of the germ layer to get corn grits. The germ can be further used as cattle feed.

Wet milling

For the wet milling of maize kernels, a moisture percentage of 15 to 16% should be maintained. Dried maize kernels should be avoided as it affects the oil quality of germ. In this milling process, moisture (45%) is added to the maize kernels followed by germ separation, then the grinding of grains followed by the hull recovery, and separation of starch and gluten from the grains.

Milling of millets

Millets are a group of cereal grasses that vary in sizes and form a part of cereal apart from the major staples such as wheat, rice, and maize. Millets serve as an important food source for people in underdeveloped countries due to their capability to grow under unfavorable climate conditions such as restricted rainfall areas. Millets are exceptionally rich in nutrients like polyphenols, dietary fiber, protein (rich in essential amino acids), and calcium in contrast to other cereals (Devi et al., 2011). Milling of millets is generally done to remove the germ and bran layer which are highly rich in phytochemicals and fiber. Amongst all the different millet varieties, Pearl millet (*Pennisetum glaucum*), Finger Millet (*Eleusine coracana*), Foxtail millet (*Setaria italica*), and Proso millet or white millet (*Panicum miliaceum*) are the major ones (Yang et al., 2012). Other minor millets include Kodo millet (*Paspalum scrobiculatum*), Teff (*Eragrostis tef*), Barnyard millet (*Echinochloa* sp.), etc. (ICRISAT, 2007; FAO, 2009; Adekunle, 2012). Milling of millets mainly involves two basic operations such as decortication or dehulling (to remove husk) and milling (to convert grains to flour).

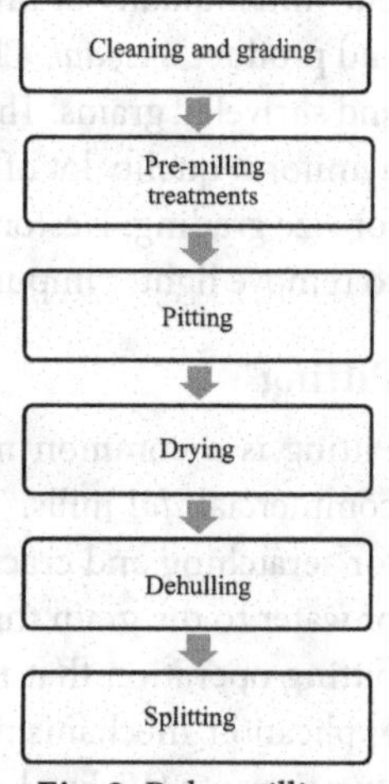

Fig. 3: Pulse milling operations

Dehulling also known as pearling is done to remove the outer layer (pericarp) from the millet grain to increase its texture, color and cooking quality (Liu et al., 2012). Decorticated millet grains can be cooked easily in less time with a softer texture, which is difficult to obtain without dehulling (Shobana et al.,

2007). Decortication extent is an important parameter of consideration and must be extended only up to 12-30% beyond this, loss of fat, ash, and fiber occurs. On a whole, the dehulling or decortications process facilitates the removal of phytic acid, polyphenols, and amylase inhibitors from the grain by increasing its protein content, starch digestibility, and the availability of minerals (ElShazali et al., 2011; Malleshi and Klopfenstein, 1998). Milling of millet grains to flour involves use of impact milling equipment such as a hammer or roller mills amongst which hammer mill gives larger size particles that are not so uniform.

Milling of pulses

Pulses form an important part of Indian diets facilitating nutrition security. Pulses are a rich source of protein from the starting of the civilization era and cereal diets are supplemented to improve their protein nutritive value. In the case of pulses, milling is mainly involved removal of the seed coat which results in the reduction of cooking time, improved digestibility, storability, and palatability (Vishwakarma et al., 2018). Dehulling reduces almost 80-90% of antinutritional compounds present in the seed coat (Singh, 1993). Another important pulse milling operation is the fine grinding of dehulled seed which separates both the protein-rich/ starch-rich fractions (Vishwakarma et al., 2018). The basic unit operations of pulse milling are depicted in Figure 3 (Verma, 2018).

Cleaning and grading

Pre-cleaning and grading of raw material to remove dust particles, dirt and foreign material, off-sized particles, immature and damaged grains is a necessary step in milling. At the initial stage, the raw material is passed through a cleaner-cum grader. The initial quality of raw material in a milling operation determines the quality of the end product, i.e., *dal.* Cleaners are used to remove impurities, foreign matter, damaged and shriveled grains. The grading is done based on the size, shape, and gravity to obtain a uniform quality lot of grains. Commonly reel machines are used at commercial mills for size grading. Destoner is used to eliminate stones and pebbles. Aspirators are used to remove lighter impurities are removed using fans or pneumatic separator

Pitting

Pitting is a common milling operation performed using an emery-coated roller in commercial *dal* mills. The emery coated rollers are utilized for their abrasive action for scratching and cracking of seed coat. Cracked seed coat facilitates the entry of oil or water to the grain throughout the pre-milling process increases its efficiency for the pitting operation that needs to be explored. These parameters may include the force application mechanism, clearance between roller and concave, roller speed, pitting duration, etc (Vishwakarma et al., 2018).

Pre-milling treatments

Pre-milling treatments are required to facilitate the separation of seed coat by diminishing the strength of its bond with cotyledons. These pre-treatments can be categorized as wet or dry pre- Wet milling is adopted at small scale milling whereas dry pre-treatment is commonly practiced for commercial milling. Wet pre-treatment process involves the soaking of the seeds in water for 6-8 hrs followed by mixing it with a slurry of 3% red earth then sun drying the mixture for 2-4 days followed by tempering of the seeds for 8-12 hrs and then separating the red earth (Vishwakarma et al., 2018). Whereas, wet pre-treatment process with the edible oils involves the pitting of the seeds (at 10–12% moisture content) followed by the application of edible oil at 0.1–1% (w/w) and then sun drying it for 2-5 days in a thin layer. After that a desired amount of water is added to increase the moisture content by 2–5%, then tempering is done for 12-18 hrs in the heaped form and then sundried to attain 10–11% moisture content, and milling (Vishwakarma et al., 2018).

Drying

Sun drying of treated and tempered grains is quite common in commercial mills but has been replaced with mechanical dryers to facilitate round the year operation. Alternate drying during day time (heating) and heaping in nights (cooling) for 2-3 days facilitate the loosening of the seed coat. The Sun-drying process is time-consuming and labor-intensive operation whereas, mechanical dryers take only 2-3 hours, thus, saves both time and energy.

Dehusking and splitting

After completion pre-treatment, grains are projected towards emery rollers where abrasive surfaces result in the removal of husk. The emery coating also referred to as carborundum which is made up of silicon carbide. The grit size of carborundum influence dehusking efficiency. The majority of mills utilize emery rollers for dehusking and splitting. Some millers use disk sheller for the splitting of dehusked grains (*gota*), whereas dehusking is performed in emery rollers only. Milled fractions *viz.* unhusked whole, dehusked whole, unhusked *dal*, dehusked *dal* and broken are separated to achieve a quality product. The by-product from the milling industry (a mixture of husk and powder) commonly goes for cattle feeding.

Factors affecting pulse milling

A range of factors like size and shape of grain, variety of cultivar, the hardness of grain, moisture content, and storage period influence the dehulling characteristics of pulses depending upon the individuality of grain and the amount of gum present between the outer hull and cotyledons (Wood and Malcolmson, 2011). Further, moisture content of seeds influences hardness, the strength of cotyledons and bond among them, adhesiveness and frictional properties of seeds (Phirke and Bhole, 1999; Wood and Malcolmson, 2011; Vishwakarma et al., 2012). For the dehulling of pulses, the optimal moisture content varies from 8 to 11% (Tiwari et al., 2007, 2010; Sreerama et al., 2009; Joyner and Yadav, 2015b). The presence of gummy substances between seed coat and cotyledons makes husk removal difficult during milling. According to the quality and amount of gums present, pulses are categorized as (i) easy to mill pulses such as chickpea, field pea, lentil, etc.; (ii) difficult to mill pulses like red gram, urad bean, mung bean,etc. (Verma, 2018).

Methods of pulse milling

The most traditional method used for milling of pulses was hand pounding, employing stones or mortar and pestle followed by the use of quern. The most primitive type was a saddle quern that consists of two stones: (i) small, smooth fist-sized stone; (ii) large, gritty base stone. Rough splits could be used to crack the seed coat, or reduced to the flour with the continued pounding. Afterward, the saddle quern changed into either a rotary or an oscillatory quern, which is also known as *chakki* consisting of two large size abrasive stones, with a slightly convex bottom and concave upper stone. In the middle of the top stone, there is a hole, occupied with the seed and is rotated using a wooden handle. Due to rotary action, the seed falls among the two large abrasive stones giving dehulled and split grains. These grains could be added with edible oil like linseed oil to impart shine or could be subjected to polishing in a cone-type polisher. Further revolving the stone reduces the split seeds into flour (Singh et al., 1992, 2000).

Due to the advancement of newer technologies, traditional techniques like stone mills are not in use because they lead to damage of grains (20-50%) and yields inferior product quality (Ali, 2004). Several machines have been developed to overcome these constraints amongst which the emery coated cylinder concave system is the most common (Singh, 1995; Tiwari and Singh, 2012). The abrasive action exhibited by the roller surface of this dehulling machine is due to the coating of carborundum emery powder on the cylindrical mild steel rollers. Carborundum is a crystal of aluminum oxide chromium, titanium, and iron traces whereas emery is impure carborundum which comprises aluminum oxide along with spinel minerals like magnetite, hercynite, and titania which are rich in iron. In dehulling machines, 14-16 to 36-40 mesh size emery grits are used. Fine mesh size emery grits are used for pitting and polishing of grains whereas large mesh size emery grits are used for the dehulling of grains. The rollers are placed at an angle of 14° inclination having a length of 0.6-0.9 m and a diameter of 0.2-0.35 m with a rotational speed of 600-1000 rpm. For the production of *dhal*, the dehulling of pulses is done by using the abrasive machine. The moisture content of about 10% of grains is required for the dehulling of pulses.

Pulse flour milling

Milling of pulses yields in flour which is used as an important ingredient for the preparation of various traditional food products like *roti, dhokla, boondi, dabo, pakora, bhujia, shiro, akara* (Wood and Malcolmson, 2011). Milling of pulses into flour is done by grinding of the dehulled seeds/*dhal* for their size reduction into fine particles. The uniform particle size of pulse flour is obtained by passing the pulse flour through a series of sieves. This process of grinding and sieving of pulses may be coupled with other operations such as dehulling and splitting of pulses (Vishwakarma et al., 2018). The flour obtained may be different because of the removal of seed coat before milling results in different pulse flour which is most commonly used and the other is milling of pulses with their seed coat which results in wholemeal pulse flour which is less commonly used. Wholemeal pulse flour can be obtained from *Kabuli* variety of chickpea which comprises a thin and opaque white seed coat that is hard to remove. Thus, the wholemeal pulse flours are rich in fiber as compared to dehulled seeds/*dhal* flour.

Pulse flour from dry kernels which are not pre-conditioned is produced by impact mills such as hammer mills. In this type of mills, a steel drum is attached with a rotating shaft (attached vertically or horizontally) that is fitted with the hammer bars which are spun inside the steel drum. When the seeds are fed into the mill the size reduction of seeds to flour takes place by exerting an impact force. Thus, the size reduction is achieved by a series of impact force. To obtain the flour of desired particle size the metal screen is changed frequently. The functional and physicochemical properties of pulse flour are affected by factors such as type of mill for the milling of pulses into flours, conditions of milling, and selection of sieves for the separation of fine size particles. Pulse flour specifications are also important for the intended end use of pulse flour as similar to the wheat flour specifications (Wood and Malcolmson, 2011).

Effect of milling operations on nutritional quality

Cereal grains and pulses are some of the important staple foods and are an abundant source of nutrients. These grains are further processed to obtain end products used for the industrial application. The other processing steps involve dehulling, milling, refining, and polishing of grains which may affect the nutritional properties of the end product by altering it. These processing steps also modify the matrices in which the nutrients are embedded in a grain, which results in influencing the nutritional property of the grains (Oghbaei and Prakash, 2016).

Effect on carbohydrates

Carbohydrates form a major fraction of cereals and pulses and act as the main source of energy in the human body. Starch content of the flour is mainly affected by the process of milling and the particle size of flour, as the screen size used in the milling process is decreased, the starch content increases (Kerr et al., 2000). The reason for this could be the separation of fiber portion during sieving. Bran consists of a large amount of insoluble dietary fiber and anti-nutrients such as phytate and tannin which binds enzymes and proteins resulting in their reduced activity. Dehulling of grains results in decreasing the significant amount of the dietary fiber (soluble and insoluble), tannin, and phytic acid content present in the grains (Ghavidel and Prakash, 2007).

Effect on protein

For the low-income group people, the major sources of protein are cereals and pulses. Besides protein quality, the other two important factors that must be met are the content of protein and its digestibility. Outer layers of the grain are rich in polyphenol and phytate that bind the minerals which are essential as cofactors, and interfere in the metabolic processes required for the utilization of protein (Landete, 2012). Hence, the removal of outer husk boosts the digestibility of grains.

Effect on minerals

Milling is a crucial process when we talk about the minerals as it greatly affects their concentrations in grains and products developed using them. The outer layer of the kernel consists of the aleurone layer and the germ layer, rich in mineral content in comparison to the starchy endosperm. The conventional milling process reduces the mineral content in flour and yields the milling residue concentrates (Oghbaei and Prakash, 2016). The disparity in the mineral content may also exist even among the outer and the inner endosperm (Brondi et al., 1984). In determining the loss in mineral content, the shape and texture of grain, as well as the technical factors of milling such as extraction rate, are important.

Technological evaluation in milling

The major objective of milling of cereals, pulses, and oilseeds are to extract the desired quality product. Technological evaluation of the milling process, thus, plays an important role in obtaining the desired quality product. For this, the automated milling units are needed with a process control system. The ultimate aim for this technological evaluation is to get a product that stands on all the quality aspects viz., particle size, color, protein content, starch damage. The monitoring of each process of milling should be monitored for effective evaluation. This can be done by using various methods shown in Figure 4.

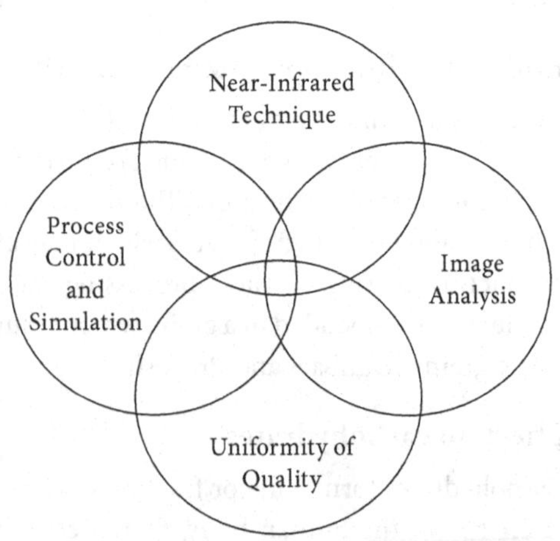

Fig. 4: Different methods for milling technological evaluation

Near-infrared technique

Near-Infrared technology was firstly adopted in 1975 by the Canadian Grain Commission for measuring the protein content in wheat grain. In 1982, the American Association of Cereal Chemists approved the determination of protein content in wheat grains using the near-infrared technique. Near-Infrared (NIR) is a spectroscopy technique that utilizes the electromagnetic spectra (1100-2500 nm) of the near-infrared region. This technique is used to measure the technological evaluation of the cereal milling process. Firstly, the incident energy directed to the sample and the organic chemistry of the sample is reflected at each wavelength, and the spectrum is noted. This technique is employed to measure the quality of cereals and cereal-based products. The advantages of using the near-infrared technology for grain analysis were fast analysis, a simple operating system, less or limited use of wet chemical analysis, and multi-component analysis. Some of the examples are to determine the wheat grain hardness (Faměra et al., 2004).

Jirsa et al. (2007) reported the use of near-infrared techniques in the quality assessment of bread. The near-infrared technique provides real-time monitoring of the milling process so that corrective action can be taken against various quality parameters like protein content, ash content, moisture content. This real-time monitoring of the process leads to a cheaper milling process with absolute product consistency. The merit of this technology is that it can be easily calibrated. Near-Infrared employs a diode array which is used as a detector. The near-infrared technique has been reported for technological evaluation of wheat milling including its physical traits, moisture and protein content, grain hardness and falling number and overall quality assessment. The use of Near-Infrared techniques for cereals quality evaluation started decades ago. The protein content of wheat and oat grains and in pulses is reported. The prediction result of the Near Infrared resembles remuneration in terms of cost, sample preparation and applicability of milled products. For the routine analysis of cereal grains, Near-Infrared techniques are preferable over traditional wet chemistry methods. Grain hardness can be easily measured which leads to the uniform grinding of the grains resulting in better prediction. The limitation is for the analysis of large sample the grinding is time-consuming.

Studies on the cereal grains properties such as protein content, density, vitreousness, and hardness have been reported using the near-infrared technique. Near-Infrared spectroscopy has been reported for the technological quality assessment of several grains like wheat, maize, barley, and oats. In barley, nitrogen and moisture content were analyzed using the near-infrared technique (Hasley, 1987). Fox et al. (2002) determined the protein content in barley and malt using near-infrared at spectral range of 1100-2500 nm. Tarr et al. (2012) observed the technological quality of malted barley by monitoring the barley grain before and after malt processing. Quality parameters that were analyzed were its nitrogen content, friability, malt extract viscosity, and β-glucan content. Not long ago, a study on the total phenolics content and *p*-coumaric acid content in barley grain were done using the near-infrared were reported by Han et al. (2017). In the study, 130 barley genotype was used and analyzed at the spectrum of 1100-2500 nm and the R^2 value was observed more than 0.93 for both the calibrated and predicted data.

The near-infrared technique is used to evaluate the heat damage in wheat grains which leads to protein denaturation and ultimately affects the processing quality. Wang et al. (2001) assessed the heat damage of wheat grains using near-infrared at spectral range of 400-1700 nm. Near-Infrared has been reported for the quality evaluation of rice during its milling to detect the moisture and protein content (Caporaso et al., 2018). Kawamura et al. (2003) studied the scanning of 33 wavelengths at the spectral range of 825-1075 nm to assess the quality of rice. The moisture content, average protein content, and sound whole kernel ratio were determined. The accuracy for protein content was $R^2 = 0.70$ for whole rice kernel and $R^2 = 0.76$ for milled rice.

The presence of microbial toxins is a concern related to the quality of the product. The near-infrared technique is also been reported for the detection of mycotoxins in the cereal grains. Berardo et al. (2005) reported the detection of mycotoxin in the maize kernel. Several mycotoxins were detected using HSI were Aflatoxin B_1 in maize kernel (Wang et al., 2015), mildew growth in wheat grains (Wang et al., 2015), *Fusarium* wilt in wheat grains (Shahin et al., 2012), toxigenic fungi in maize kernel (Fiore et al., 2010), aflatoxigenic fungal infection in rice kernels (Sirisomboon et al., 2013).

Hyperspectral Imaging (HSI)

Hyperspectral imaging (HSI) is a new technology used for the technological evaluation of the milling process. This technique involves both spectroscopy as well as the imaging process for quality measurements such as chemical properties and spatial distribution. Hyperspectral imaging collects the data of the spatial distribution of images in three-dimension, known as Hypercube. Hyperspectral imaging has been used as a versatile device for the technological evaluation of milling properties. The main advantage is that it can be used to assess the grain individually rather than taking average data. Other advantages are multi-component analysis, assessment of mechanism involved in the germination process, quantification of complex properties like moisture and protein content, prediction of rheological properties of products like bread by imaging wheat grains.

Various researches have been done for the modernization of technological evaluation of milling. Hyperspectral imaging is one of the techniques which have been used for the cereal grain evaluation. Recently, research was conducted for the speedy evaluation of the total hull-less content of the oat grains (Serranti et al., 2013). Hyperspectral imaging technique has been shown its potential over other techniques like near-infrared spectroscopy for the assessment of intact grains. Hyperspectral imaging techniques have several various applications like the detection of extraneous material and impurities in the grain and the final milled product. Impurities like the presence of other seeds and animal excreta can be detected. Vitreousness in wheat grain was determined using Hyperspectral imaging at 650-1100 nm spectrum (Gorretta et al., 2006). Shahin et al. (2008) conducted the same study on the Canadian wheat in which the grains were scanned at a spectral range of 950-2500 nm resulting in differentiation of wheat grains into vitreous and starchy grains. Hyperspectral imaging is also used for the quantitative assessment of the cereal grains as it offers a great advantage to determine grains having a different composition by scanning multiple grains per time. Deng et al. (2014) employHyperspectral imaging for the classification of rice grains using 700 rice kernels from 6 short rice grain varieties, and the highest accuracy observed was 91.95%.

During the germination and sprouting of cereal grains, the enzymatic activity of grains is enhanced which thus affects the milling quality of the cereal grains. For example, α-amylase is one of the important enzymes present in most of the cereal grains. Research by Koç et al. (2008) reveals that the Hyperspectral imaging was able to analyze the alteration in pre-germinated and germinated wheat grains at a spectral range of 1000-1700 nm using an indium gallium arsenide (InGaAs) focal plane array detector. Hyperspectral imaging has been reported to assess the α-amylase activity in wheat grains using a short wavelength infrared hyperspectral imaging system (SWIR-HSI) (Xing et al., 2009). Caporaso et al. (2017) utilize Hagberg falling number (HFN) as the reference measurement for estimation of sprout damage in wheat grain. Hyperspectral imaging technique at 900-1700 nm spectrum has been utilized to analyze the pre-germination quality of barley grain to assess its viability and germination activity Arngren et al., 2011).

Hyperspectral imaging technique is reported to detect adulteration in cereal grains. Sun et al. (2014) observed that Hyperspectral imaging at 390-1050 nm spectrum detects the adulteration using a support vector machine in the rice grains. The predicted accuracy obtained was 98%. Rice quality parameters like starch gelatinization, cooking properties, protein, and amylase content cooking time, cooking hardness, and stickiness were also assessed by using a hyperspectral imaging system.

Hyperspectral imaging technique for the technological evaluation of maize kernels has been studied by several researchers. Weinstock et al. (2006) reported the assessment of oil content and oleic acid concentration in a single maize kernel using HSI at 950-1700 nm spectrum. Wang et al. (2015) reported the use of the HSI technique for the differentiation of waxy and sweet corn. Hypercubes were acquired at a spectral range of 400-1000 nm. The prediction accuracy obtained was 90.7%. Other quality parameters of maize kernels like hardness were also studied.

Simulation

The milling process is said to be efficient based on their productivity and product quality. The simulation and mathematical modeling is another aspect of the technological evaluation of the milling. Simulation process with a focus on computer-aided manufacturing impacts the properties of milling machines (Brecher et al., 2017). The simulation process is based on the geometric description of the milling machine resulting in the overall process control. The simulation process also assesses the energy use for the milling process. Energy use is considered as a performance index during the process planning and process control. For this, the simulation approach evaluates the energy used is milling operations. The quality standards for this simulation approach are as per ISO standards which involve the mechanistic and energy model of the machine. The simulation approach is further validated by comparing the results with the predicted value.

Quality evaluation

Quality indices are one of the most important aspects to evaluate the quality of the milled products. Quality parameters which are considered for the milling of cereal grains are test weight of kernel, color, moisture content, percentage of damaged kernels cleanliness, plumpness, and soundness (Lásztity, 2009).

Image analysis

To detect the different admixtures with foreign matter in cereal grains, a digital image analysis system is used. Image analysis of the cereal grains also helps to maintain the uniformity of the grains resulting in uniform particle size. Image analysis of cereal grains is a core of Computer vision. Computer vision is a process that focuses on acquiring imaging-based automatic inspection and process control (Saini et al., 2014). In other words, computer vision is defined as a process to analyze the image and collecting the data for controlling the process (Relf, 2003). Image analysis of cereal grains determines the size of grains which is related to grain quality, the color of the grains. The color imaging analysis of the grains helps to differentiate the poor quality grains by the difference in the grain color. This technique determines the morphological characteristics of the grains by employing a neural network. Image analysis consists of a chain of image operations resulting in increased image quality by determining the level of processing viz., low, intermediate, and high. For the image analysis of cereal grains, GrainCheck 310 (FOSS Tecator, Höganäs, Sweden) is used. This image analyzer is based on an artificial neural network (ANN) system that estimates the grain quality by extracting the data of the shape and color of the grains.

Thus, the image analysis technique is a core analysis method for the technological evaluation of the milling properties of grains. This technique focuses on the morphology shape, and texture of grain which are an important aspect to determine the quality of grains. The image analysis technique is also employed to detect the presence of infestation in the grains and to classify the cereal grains in different classes.

Automation in the milling industry

The utilization of Computer Integrated Manufacturing (CIM) and Programmable Automation (PA) in the milling industries offers frontiers in terms of enhancing quality and production control. Improving control results in energy saving and improving the product consistency by stabling the important process variables, to allow easy margins to be lessened. To accomplish these purposes advanced technologies are needed. Highly integrated processes like mill processes that act as an interacting element during the milling are difficult to control by utilizing basic control units. Furthermore, the use of simple controller units results in a delay between the action of the controller and its response (both system response as well as non-linear response) from the milling process. Thus, a highly equipped controller is needed which then makes the milling process flexible, for example, the process plant may produce different end products depending upon the operations at different throughputs. Process control can be categorized into two groups: (i) conventional process control; (ii) advanced process control (Owens, 2001).

The conventional process controls include Single-Input Single-Output Proportional Integral Differential (SISO-PID) controller, which has been in use for several years. In this controller, the process dynamics are reasonably constant which possesses minimal lag in the process. The main objective of this controller is to adhere to some measurable set point. Feedforward control, Ratio control, Constraint control, etc. are the few conventional controls. This type of controllers is employed in various milling operations like tempering (addition of water into wheat grains) and the addition of ingredients into the flour. However, the advanced process control system is measured when the other process control approach i.e., conventional process control is unable to develop adequate performance. The advanced process control system involves Model-based predictive control like Rule-based control, Fuzzy control, and Adaptive control that are linked with some systems like PLC (Programmable Logic Controller) DCS (Distributed Control System). This system helps in decision making in terms of the optimum solution for the process requirements and condition as proposed by an advanced process control system.

Conclusion

Foodgrains is a principal constituent of the human diet for years and play an important part in ensuring nutritional security amongst the population. For the daily survival of billions of people worldwide, a major extent for grains like rice, wheat, maize, and pulses and to a lesser area, grains such as sorghum and millets are a chief staple food. These food grains are going through primary processing which allows their further utilization for the manufacture of products or cooking purposes. Milling operations involve the application of different mechanical forces to overcome the interior bonding forces bringing transformation from the state of solid grains to different desirable forms such as flour or meal depending on the desired end-use. Technological evaluation of various milling operations is necessary to establish a conclusion regarding the performance of operation to promote better efficiency in operations. It involves applications of a series of principles, techniques, tools, and methods which efficiently assess the significant importance of the technology and its involvement in enhancing the efficiency of milling operations. The ultimate aim for this technological evaluation is to get a product that stands on all the quality aspects viz., particle size, color, protein content, starch damage.

This chapter has discussed the various ways in which the milling industry has been evolved in terms of technology. The development of the improved capacity of machines used in milling industries is increasing dramatically but the properties of the material being used are a limiting factor. Adoption of process control employing computers also assists the longer operational runs in plants as well as minimizing downtime is an important step. To wind up, the technological evaluation of milling processes leads to the incremental development of milling industries in the future with the introduction of newer and innovative technologies for efficiency enhancement.

References

Adekunle, A.A. (2012). Agricultural innovation in sub-saharanafrica: experiences from multiplestakeholder approaches. Forum for Agricultural Research in Africa, Ghana. ISBN 978-9988- 8373-2-4.

Ali, N. (2004). Postharvest technology and value addition in pulses. In: Pulses in a New Perspective, pp. 530–543. Ali, M. Singh, B. B. Kumar, S. and Dhar, V., Eds., Indian Society of Pulses Research and Development, Indian Institute of Pulse Research, Kanpur, India.

Anthony, J. W. Bideaux, R. A. Bladh, K. W. & Nichols, M. C. (1997). Corundum. In: Handbook of Mineralogy: III (Halides, Hydroxides, Oxides), pp. 1–628. Chantilly, VA, US: Mineralogical Society of America.

Arngren, M., Hansen, P.W., Eriksen, B., Larsen, J., & Larsen, R. (2011) Analysis of pregerminated barley using hyperspectral image analysis. *Journal of Agricultural and Food Chemistry* 59: 11385–11394.

Berardo, N., Pisacane, V., Battilani, P., Scandolara, A., Pietri, A.,& Marocco, A. (2005) Rapid detection of kernel rots and mycotoxins in maize by near-infrared reflectance spectroscopy. *Journal of Agricultural and Food Chemistry* 53: 8128–8134.

Brecher, C., Wellmann, F., & Epple, A. (2017). Quality-predictive CAM simulation for NC milling. *Procedia Manufacturing*, 11, 1519-1527.

Brondi, M., Ciardi, A., & Cubadda, R. (1984). Transfer of trace elements from the environment to food chain levels in grains and their products. La Rivista della SocietaÁ Italiana di Scienza

Caporaso, N., Whitworth, M.B., & Fisk, I.D. (2018). Near-Infrared spectroscopy and hyperspectral imaging for non-destructive quality assessment of cereal grains. *Applied Spectroscopy Reviews*, 53(8), 667-687.

Caporaso, N., Whitworth, M.B., & Fisk, I.D. (2017) Application of calibrations to hyperspectral images of food grains: Example for wheat falling number. *Journal of Spectral Imaging* 6: 1–15.

Caporaso, N., Whitworth, M.B.,& Fisk, I.D. (2017) Protein content prediction in single wheat kernels using hyperspectral imaging. *Food Chemistry* 240: 32–42.

DelFiore, A., Reverberi, M., Ricelli, A., Pinzari, F., Serranti, S., Fabbri, A., Bonifazi, G.,& Fanelli, C. (2010). Early detection of toxigenic fungi on maize by hyperspectral imaging analysis. *International Journal of Food Microbiology* 144: 64–71.

Deng, X., Zhu, Q., & Huang, M. Semi-supervised classification of rice seed based on hyperspectral imaging technology. American Society of Agricultural and Biological Engineers. 141912601.

Deshpande, S. D. Balasubramanya, R. H. Khan, S. & Bhat, D. K. (2007). Influence of premilling treatments on dal recovery and cooking characteristics of pigeon pea. *Journal of Agricultural Engineering* 44(1):53–57.

ElShazali, A.M, Nahid, A.A., Salma, H.A. & Elfadil, E.B. (2011). Effect of radiation process on antinutrients, protein digestibility and sensory quality of pearl millet flour during processing and storage. *International Food Research Journal*, 18 (4): 1401- 1407.

Emami, S. S. T. (2007). Hydrocyclone fractionation of chickpea flour and measurement of physical and functional properties of flour and starch and protein fractions. Ph.D. Thesis, Department of Agricultural and Bioresource Engineering, University of Saskatchewan, Saskatoon, Canada.

Enrique, Ortega-Rivas (2012). Food Engineering Series; Non-Thermal Food Engineering Operation; Size Reduction Springer Science & Business Media LLC, pp 71-87.

Famĕra, O., Hrušková, M., and Novotná, D. (2004). Evaluation of methods for wheat grain hardness determination. *Plant, Soil and Environment*, 50, 489-493.

FAO (2011). Food and Agriculture Organization of the United Nations. Processing and storage of food grains by rural families. Agricultural Services Bulletin No. 53. Rome – Italy.

FAO. (2009). FAOSTAT. Food and Agriculture Organisation of the United Nations. FAOSTAT. http://faostat.fao.org/site/339/ default.aspx

Fox, G., Onley-Watson, K., & Osman, A. (2002). Multiple linear regression calibrations for barley and malt protein based on the spectra of hordein. *Journal of the Institute of Brewing* 108: 155–159.

Ghavidel, R., & Prakash, J. (2007). The impact of germination and dehulling on nutrients, antinutrients, in vitro iron and calcium bioavailability and *in vitro* starch and protein digestibility of some legume seeds. *LWT-Food Science and Technology*, 40, 1292–1299.

Gorretta, N., Roger, J. M., Aubert, M., Bellon-Maurel, V., Campan, F., & Roumet, P. (2006). Determining vitreousness of durum wheat kernels using near infrared hyperspectral imaging. *Journal of Near Infrared Spectroscopy*, 14(4), 231-239.

Halsey, S.A. (1987). Analysis of whole barley kernels using near-infrared reflectance spectroscopy. *Journal of the Institute of Brewing* 93: 461–464.

Han, Z., Cai, S., Zhang, X., Qian, Q., Huang, Y., Dai, F., & Zhang, G. (2017) Development of predictive models for total phenolics and free p-coumaric acid contents in barley grain by near-infrared spectroscopy. *Food Chemistry* 227: 342–348.

ICRISAT. (2007). International Crops Research Institute for the Semi-Arid Tropics. Annualreport. http://test1.icrisat.org/Publications/EBooksOnlinePublications/AnnualReport-2007.pdf

Indira, T.N., Bhattacharya, S. (2006). Grinding characteristics of some legumes. *Journal of Food Engineering* 76, 113–118.

Joyner, J. J. & Yadav, B. K. (2015b). Optimization of continuous hydrothermal treatment for improving the dehulling of black gram (*Vigna mungo* L). *Journal of Food Science and Technology* 52(12):7817–7827.

Kaul, RN. & Egbo, CO. (1985). Introduction to Agricultural Mechanization. Publisher, Macmillian Education, Limited.

Kawamura, S., Natsuga, M., Takekura, K., & Itoh, K. (2003) Development of an automatic rice: Quality inspection system. *Computers and Electronics in Agriculture* 40: 115–126.

Kawuyo, U. A., Lawal, A. A., Abdulkadir, J. A., & Dauda, Z. A. (2017). Performance Evaluation of a Developed Grain Milling Machine. *Arid Zone Journal of Engineering, Technology and Environment*, 13(1), 1-5.

Kerr, W., Ward, C., McWatters, K., & Resurreccion, A. (2000). Effect of milling and particle size on functionality and physicochemical properties of cowpea flour. *Cereal Chemistry*, 77, 213–219.

Koç, H., Smail, V.W., & Wetzel, D.L. (2008) Reliability of InGaAs focal plane array imaging of wheat germination at early stages. *Journal of Cereal Science* 48: 394–400.

Kulkarni, S. D. (1989). Pulse processing machinery in India. Agric. Mech. Asia Afr. Latin Am. 20(2):42–48.

Lal, R. R. and Verma, P. (2007). Post Harvest Management of Pulses. Indian Institute of Pulse Research, Kanpur, India.

Landete, J. M. (2012). Updated knowledge about polyphenols: Functions, bioavailability, metabolism, and health. *Critical Reviews in Food Science and Nutrition*, 52, 936–948.

Lásztity, R. (2009). Food Quality and Standards. Eolss Publishers Company Limited.

Liu, J., Tang, X., Zhang, Y. & Zhao, W. (2012). Determination of the volatile composition in brown millet, milled millet and millet bran by gas chromatography/ mass spectrometry. *Molecules* 17: 2271–82.

Malleshi, N.G. & Klopfenstein, C.F. (1998). Nutrient composition and amino acid contents of malted sorghum, pearl millet and finger millet and their milling fractions. *Journal of Food Science and Technology*, 35 (3): 247-249.

Mijinyawa, Y., Ajav, E.A., Ogedengbe. K.O. and Aremu, A.K (2007). The Agricultural Engineering in Introduction to Agricultural Engineering. Aluelemhegbe Publishers, Ibadan, (ISBN 978 9780847296) Pages 1 – 39.

Nagah, A., & Seal, C. (2005). *In vitro* procedure to predict apparent antioxidant release from wholegrain foods measured using three different analytical methods. *Journal of the Science of Food and Agriculture*, 85, 1177–1185.

Narasimha, H. V. Ramakrishnaiah, N. & Pratape, V. M. (2003). Milling of pulses. In: Handbook of Post Harvest Technology, pp. 427–454.Chakraverty, A. Mujumdar, A. S. Raghavan, G. S. V. and Ramaswamy, H. S., Eds., Marcel Dekker, Inc., New York, USA.

Oghbaei, M., & Prakash, J. (2016). Effect of primary processing of cereals and legumes on its nutritional quality: A comprehensive review. *Cogent Food & Agriculture*, 2(1), 1136015.

Ogwuagwu, V.O (2007). Simulation Study of the Fracture Characteristics of Locally Fabricated Aluminium Bar used as Connecting Rod in Vibratory Sieves, *Leonardo Journal of Sciences* ISSN 1583-0233, Issue 10, January-June 2007 pp. 93-100

Owens, G. (Ed.). (2001). Cereals Processing Technology. CRC Press.

Parada, J., & Aguilera, J. (2007). Food microstructure affects the bioavailability of several nutrients. *Journal of Food Science*, 72, R21–R32.

Phirke, P. S. & Bhole, N. G. (1999). The effect of pre-treatment on the strength and dehulling properties of pigeonpea grain. *International Journal of Food Science and Technology*. 34(2):107–113.

Prom-u-thai, C., Huang, L., Glahn, R., Welch, R., Fukai, S., & Rerkasem, B. (2006). Iron (Fe) bioavailability and the distribution of anti-Fe nutrition biochemicals in the unpolished, polished grain and bran fraction of five rice genotypes. *Journal of the Science of Food and Agriculture*.

Ryan J.G. & Spenser, D.C. (2008). Future Challenges and Opportunities for Agriculture in the Semi-Arid Tropics. Andrhraprades; India: International Crops Research Institute for the SAT.

Sanjeewa, T.W.G., Wanasundara, J.P.D., Pietrasik, Z., & Shand, P.J. (2010). Characterization of chickpea (Cicer arietinum L.) flours and application in low-fat pork bologna as a model system. *Food Research International* 43, 617–626.

Serranti, S., Cesare, D., Marini, F., & Bonifazi, G. (2013) Classification of oat and groat kernels using NIR hyperspectral imaging. *Talanta* 103: 276–284.

Shahin, M.A., & Symons, S.J. (2008) Detection of hard vitreous and starchy kernels in amber durum wheat samples using hyperspectral imaging (GRL Number M306). NIR News 19: 16–18.

Shahin, M.A., & Symons, S.J. (2012). Detection of fusarium damage in Canadian wheat using visible/near-infrared hyperspectral imaging. *Journal of Food Measurement and Characterization* 6: 3–11.

Shahin, M.A., Hatcher, D.W., & Symons, S.J. (2010). Assessment of mildew levels in wheat samples based on spectral characteristics of bulk grains. *Quality Assurance and Safety of Crops and Foods* 2: 133–140.

Shahzadi, N., Butt, M.S., Rehman, S.U., Sharif, K., (2005). Rheological and baking performance of composite flours. *International Journal of Agriculture and Biology* 7, 100–104.

Shobana, S. & N. G. Malleshi. (2007). Preparation and functional properties of decorticated finger millet (Eleusine coracana).*Journal of Food Engineering* 79:529–538.

Singh, U. (1993). Protein quality of pigeon pea (*Cajanus cajan* L.) as influenced by seed polyphenols and cooking process. *Plant Food Human Nutr.* 43:171–179.

Singh, U. (1995). Methods for dehulling of pulses: A critical appraisal. *Journal of Food Science and Technology* 32(2):81–93.

Singh, U., Santosa, B.A.S., & Rao, P.V. (1992). Effect of dehulling methods and physical characteristics of grains on dhal yield of pigeonpea (*Cajanus cajan* L.) genotypes. *Journal of Food Science and Technology*. India 29, 350–353.

Singh, U., Williams, P.C. & Petterson, D.S., (2000). Processing and grain quality to meet market demands. In: Knight, R. (Ed.), Linking research and marketing opportunities for pulses in the 21st century. Kluwer Academic Publishers, Dordrecht, pp. 155–166.

Sirisomboon, C.D., Putthang, R., & Sirisomboon, P. (2013). Application of near infrared spectroscopy to detect aflatoxigenic fungal contamination in rice. *Food Control* 33: 207–214.

Sreerama, Y. N. Sashikala, V. B. & Pratape, V. M. (2009). Effect of enzyme pre-dehulling treatments on dehulling and cooking properties of legumes. *Journal of Food Engineering* 92:389–395.

Sun, J., Jin, X., Mao, H., Wu, X., & Yang, N. (2014) Application of hyperspectral imaging technology for detecting adulterate rice. *Transactions of the Chinese Society of Agricultural Engineering* 30: 301–307.

Tarr, A., Diepeveen, D., & Appels, R. (2012) Spectroscopic and chemical fingerprints in malted barley. *Journal of Cereal Science* 56: 268–275.

Tiwari, B. K. & Singh, N. (2012). Pulse Chemistry and Technology. The Royal Society of Chemistry, Cambridge, UK.

Tiwari, B. K. Jaganmohan, R. & Vasan, B. S. (2007). Effect of heat processing on milling of black gram and its end product quality. *Journal of Food Engineering*. 78:356–360.

Tiwari, B. K. Jaganmohan, R. Venkatachalapathy, N. Anand, M. T. Surabi, A. & Alagusundaram, K. (2010). Optimisation of hydrothermal treatment for dehulling pigeon pea. *Food Review International* 43:496–500.

Verma, P. (2018). Processing and value addition of pulses. *Anusandhan- AISECT University Journal.* 6(13): P-ISSN 2278-4187.

Vishwakarma, R. K. Shivhare, U. S. & Nanda, S. K. (2012). Predicting guar seed splitting by compression between two plates using Hertz theory of contact stresses. *Journal of Food Science* 77(9):E231–E239.

Vishwakarma, R. K., Shivhare, U. S., Gupta, R. K., Yadav, D. N., Jaiswal, A., & Prasad, P. (2018). Status of pulse milling processes and technologies: A review. *Critical Reviews in Food Science and Nutrition*, 58(10), 1615-1628.

Wang, D., Dowell, F., & Chung, D. (2001). Assessment of Heat-damaged wheat kernels using near-infrared spectroscopy. 1. *Cereal Chemistry* 78: 625–628.

Wang, L., Sun, D.W., Pu, H., & Zhu, Z. (2015) Application of hyperspectral imaging to discriminate the variety of maize seeds. *Food Analytical Methods* 9: 225–234.

Wang, W., Lawrence, K., Ni, X., Yoon, S.C., Heitschmidt, G., and Feldner, P. (2015) Near-infrared hyperspectral imaging for detecting Aflatoxin B1 of maize kernels. *Food Control* 51: 347–355.

Weinstock, B.A., Janni, J., Hagen, L., & Wright, S. (2006) Prediction of oil and oleic acid concentrations in individual corn (*Zea mays* L.) kernels using near-infrared reflectance hyperspectral imaging and multivariate analysis. *Applied Spectroscopy* 60: 9–16.

Wood, J. A. & Malcolmson, L. J. (2011). Pulse milling technologies. In: Pulse Foods: Processing, Quality and Nutraceutical Applications, pp.193–221. Tiwari, B. Gowen, A. and McKenna, B., Eds., Academic Press,UK.

8

Fermented Food Based on Cereal and Pulses

Rachna Gupta and Murlidhar Meghwal

Department of Food Science and Technology, National Institute of Food Technology, Entrepreneurship and Management, Sonipat, Haryana, India

Introduction

Fermentation is an alluring procedure of biochemical adjustment of essential nourishment microorganisms and their enzymes accomplish by lattice. Fermentation is a utilized to upgrade bio-accessibility and bioavailability of supplements from various crops and improves organoleptic properties (e.g., appearance, generation of sour taste and new aroma compounds, and textural changes such as consistency or viscosity) just as increase the shelf life of realistic usability. It makes foodstuff safe not just hindering the growth of pathogenic microorganisms because of antimicrobial growth of lactic acid and it also detoxifies aflatoxin. With these advantages, fermentation is a successful method to reduce the risk of mineral deficiency among population, particularly in developing nations where unrefined cereal or pulses are profoundly consumed. Tragically, it is additionally connected with multiplication of microorganisms, for ex., yeast and molds caused food safety concerns, (Omemu et al; 2011), decrease in provitamin A and antioxidant (Ortiz et al; 2017), carotenoids, just as loss of nutrients and minerals (Hotz & Gibson, 2007).

Fermentation is a minimum cost-efficient procedure for production and preservation of nutrient, therefore fermented food is an important piece of the eating regimen in numerous of the world. In the nineteenth century the scientific study of microbiology began, and just because, the scientific procedure of fermentation was comprehended. several classifications have been utilized to sort the assorted exhibit of the fermented nourishment produced dependent on various microorganisms, and food types, and kinds of fermentation included.

The microorganisms associated with fermentation process fall into three classes: bacteria, yeast, and molds. Yeasts are the principal microorganisms for the fermentation of breads, whereas molds is mainly used in processing of cheese and legumes. Bacteria is included only of the fermentation of cereal product. cereal and tuber fermentation are a lactic and acetic acid producing bacteria. yeast is a piece of mix microbial fermentation by bacteria in weaning food; all the more frequently, be that as it may, alcoholic fermentation of yeast may not totally alluring. Four genera of lactic acid producing bacteria are recognized as overwhelming cereal fermentation to be specific, Lactobacillus, Leuconostoc, Pediococcus, and Lactococcus spp. all of which require carbohydrate for the energy. Lactic acid and other natural acids produced from hexoses bring down the pH from about 6.5 to about 3.6. In the event that the pH is beneath 4.0, fermented product retard the growth of pathogens and along these lines ensure fermented items.

Fermentation techniques have been throughout the years improved, and novel processing technique are being investigated and executed to satisfy the needs of high-quality processed food and drinks, which would be more secure, free from manufactured synthetic chemical, and simultaneously nutritious and fresh. From farm to fork, a lot of novel techniques have been employed for the optimum satisfaction of customer and also to meet the demands of fermented food products. The cutting-edge fermented food and food industry is profoundly competitive and imaginative and is regularly during the time spent overhauling, advancement, and refinement in the innovation for quality improvement and new product advancement from various food sources.

The processing and utilization of the fermented food are not uniform all through the globe; it's advantageous to note here that the monstrously different fermented food items are as of now winning because of social assorted variety worldwide and inclination of the purchasers of a specific area. The important portion of fermented food products incorporate dairy items, beer, wine, spirits, fermented fish, meat, and vegetables. Dairy items have the most elevated stake among different component of fermented foods as milk or milk products are consumed by six billion people around the world, and it provide 6–7% of the dietary protein supply in Asia and Africa and 19% in Europe. Keeping in perspective on the wellbeing advancing activities of probiotic microbial strains, the demand of fermented dairy products as well as non-dairy items is rising quickly. Utilization of probiotics in dairy items, for example, yogurt, cheddar, desserts, and so forth are hardly any such developments that have expanded the utilization of dairy items by a wide margin as of late. Throughout the years, the fermentation processes have been refined, and fermented products have been strengthened in light of nutraceutical segments keeping of human's wellbeing and health. Various other examples such as application of encapsulation, genetic engineering, nonthermal processing for food safety and preservation, active packaging technologies, and computational intelligence techniques for forecasting problematic wine, beer, and other alcoholic beverages have been regarded as ultramodern devices and systems that have contributed colossally toward the development of food and beverage industries.

Universally fermented foods: current status and future possibilities

Universally functional food market advertise was worth USD 64,871 million out of 2018 and is relied upon to reach roughly USD 99,975 million by 2025, (www.grandviewresearch.com). The significance of milk and milk products is in an expanding pattern throughout the past three decades. Worldwide milk yield in 2018 is assessed at 843 million tons, an expansion of 2.2 percent from 2017, driven by production expansions in India, Turkey, The European Union, Pakistan, The United States of America and Argentina, yet somewhat counterbalance by decreases in China and Ukraine (http://www.fao.org). So also, worldwide beer production from 1998 to 2018. In 2018, the worldwide beer production amounted to about 1.94 billion hectolitres, up from 1.3 billion hectolitres in 1998. Beer is a generally expended drink the world over, made out of water, malt, hops and yeast as fundamental ingredients. Worldwide driving countries in beer production are China, The United States and Brazil (https://www.statista.com). Wine production arrived at a record level in 2018 arriving at 293 million hectolitres, a great recuperation from a calamitous 2017. The greatest wine producers in the world 2018 are Italy, France, Spain and the USA **(www.bkwine.com).**

Different types fermented food products from the regions of Africa, North America and South America, Asia and Oceania and Middle East and Europe.

Fermented foods of Africa

For the purpose of availability, fermented foods in Africa are classified into the following major groups- (1) Fermented non-alcoholic cereal substrate (2) Starchy root crops (3) Fermented dairy (4) Fermented Fish and Meat (5) Fermented vegetables (6) Alcoholic drinks.

Fermented foods of Asia and Oceania

The fermented food of Asia and Oceania have been categorized into five groups: (1) Fermented cereal products (2) Fermented fish and meat products (3) Fermented vegetable products (4) Fermented soybean products and (5) Alcoholic beverages.

Fermented foods of North and South America

The fermented food products of this region are categorized into five groups: (1) Fermented cereal products (2) Alcoholic beverages (3) Fermented vegetable products (4) Fermented milk products

Fermented foods products of Europe and Middle East

The fermented food products of this region are categorized into five groups: (1) Fermented cereal products (2) Fermented dairy products (3) Fermented meat products (4) Fermented vegetable products

Table 1: Commercially essential fermented foods and use of microorganism

Finished Products	Microorganisms used	Nature of fermentation	Major producer	Reference
Wine	Saccharomyces cerevisiae	Submerged	Italy, France, Spain	Mishra et al. (2017)
Beer	Saccharomyces cerevisiae, Saccharomyces pastorianus	Submerged	China, USA	Stewart (2016)
Whiskey	Saccharomyces cerevisiae	Submerged	France, Scotland, USA, Canada	Walker and Hill (2016)
Yogurts	Streptococcus thermophilus, Lactobacillus delbrueckii	Submerged	France, Ireland, Canada, USA	Han et al. (2016)
Cheese	Lactococcus, Lactobacillus, Streptococcus sp., Penicillium roqueforti	Solid-state fermentation	Germany, the Netherlands, France, USA	Mishra et al. (2017)
Acidophilus milk	Lactobacillus acidophilus	Submerged	North America, Europe, Asia	Yerlikaya (2014)
Sauerkraut	Leuconostoc sp., Lactobacillus brevis, Lactobacillus plantarum	Solid-state fermentation	Europe	Swain et al. (2014)
Fish sauce	Lactic acid bacteria (halophilic), Halobacterium salinarum, Halobacterium cutirubrum, Bacillus sp.	Submerged	Thailand, Korea, Indonesia	Lopetcharat et al. (2001)
Fermented meat	Lactobacillus sp, Micrococcus sp., Staphylococcus sp.	Solid/submerged fermentation	Europe	Holck et al. (2015)

Categorization of Fermented Foods

Fermented food is classified in a number of ways, by categories (Yokotsuka, 1982).

(1) alcoholic beverages fermented by yeasts,

(2) vinegars fermented with *Acetubacter,*

(3) milks fermented with *lactobacilli,*

(4) pickles fermented with *lactobacilli,*

(5) fish or meat fermented with *lactobacilli* and

(6) plant proteins fermented with molds with or without *lactobacilli* and yeasts; by classes (Campbell-Platt, 1987) - (1) beverages, (2) cereal products, (3) dairy products, (4) fish products, (5) fruit and vegetable products, (6) legumes and (7) meat products; by commodity (Odunfa, 1988) - (1) fermented starchy roots, (2) fermented cereals, (3) alcoholic beverages, (4) fermented vegetable proteins and (5) fermented animal protein; by commodity (Kuboye, 1985) - (1) cassava based, (2) cereal, (3) legumes and (4) beverages. Dirar (1993) states that the Sudanese traditionally classify their foods not on the basis of microorganisms or commodity but on a functional basis: (1) Kissar (staples) - porridges and breads such as aceda and kissra, (2) Milhat (sauces and relishes for the staples), (3) marayiss (30 types of opaque beer, clear beer, date wines and meads and other alcoholic drinks) and (4) Akil-munasabat (food for special occasions). Steinkraus (1996)

Classification of fermented foods in a different way

The main groups of substrates-based fermented foods are as follows:

1. Fermented milk foods
2. Fermented cereal foods
3. Fermented vegetable foods
4. Fermented soybean foods and non-soybean foods
5. Fermented meat products
6. Fermented fish products
7. Fermented root/tuber products
8. Fermented beverages and Asian amylolytic starters
9. other fermented products (fermented tea, cocoa, vinegar, *nata, pidan*, etc.)

Fermented Milks

Fermented milks are classified into two main categories based on the presence of dominant microorganisms:

(i) lactic acid fermentation which is dominated by species of Lactic acid bacteria. It is a thermophilic type (e.g., yogurt, Bulgarian buttermilk), probiotic type (acidophilus milk, yakult, bifidus milk), and the mesophilic type (e.g., natural fermented milk, cultured milk, cultured cream, cultured buttermilk).

(ii) fungal-lactic fermentations where LAB and yeasts species cooperate to generate the final product and consist of alcoholic milks (e.g., *kefir, koumiss*, acidophilus-yeast milk), and moldy milks (e.g., *viili*). fermented milk of starter culture is of two types depending on the principal function, primary cultures to participate in the acidification, and secondary cultures for flavour, aroma, and fermenting activities. The species involved as primary cultures in milk fermentation are *Lactococcus lactis* subsp. *cremoris, Lc. lactis* subsp. *lactis, Lactobacillus delbrueckii* subsp. *delbrueckii, Lb. delbrueckii* subsp. *lactis, Lb. helveticus, Leuconostoc* spp., and *Streptococcus thermophilus.*

Secondary cultures is used in cheese making are *Brevibacterium linens, Propionibacterium freudenreichii, Debaryomyces hansenii, Geotrichum candidum, Penicillium camemberti,* and *P. roqueforti* for the development of flavour and texture during the ripening of cheese.

Besides primary and secondary cultures, some nonstarter lactic acid bacteria (NSLAB) microbiota are usually present in high numbers which include *Enterococcus durans, Ent. faecium, Lactobacillus casei, Lactobacillus plantarum, Lactobacillus salivarius,* and *Staphylococcus* spp. Yogurt is a widely consumed highly nutritious fermented milk as a coagulated milk product resulting from the fermentation of milk by *Strep. thermophilus* and *Lactobacillus delbrueckii* subsp. *bulgaricus* (formerly *Lactobacillus bulgaricus*). *Lactobacillus acidophilus, Lactobacillus casei, Lactobacillus rhamnosus, Lactobacillus gasseri, Lactobacillus johnsonii,* and *Bifidobacterium* spp., are among the most common adjunct cultures in yogurt fermentation. Fermented milk products that are manufactured using starter cultures containing yeasts include acidophilus-yeast milk, *kefir, koumiss,* and *viili. Lactobacillus acidophilus, Lactobacillus. amylovorus, Lactobacillus. crispatus, Lactobacillus. gallinarum, Lactobacillus. gasseri,* and *Lactobacillus. johnsonii* are reported from acidophilus milk.

Fermented cereal foods

Cereal fermentation is characterized by a complex microbial ecosystem, mainly represented by the species of LAB and yeasts, and fermented bread has a characteristic property such as palatability and high sensory quality. The species of *Enterococcus, Lactococcus, Lactobacillus, Leuconostoc, Pediococcus, Streptococcus,* and *Weissella* are commonly associated with cereal fermentation. A native strain of *Sacch. cerevisiae* is the principal yeast of most bread fermentations. Other non-*Saccharomyces* yeasts are also significant in many cereal fermentations which include *Candida, Debaryomyces, Hansenula, Pichia, Trichosporon, Yarrowia. Lactobacillus plantarum, Lactobacillus panis, Lactobacillus sanfrancisensis, Lactobacillus pontis, Lactobacillus brevis,*

Lactobacillus curvatus, Lactobacillus sakei, Lactobacillus alimentarius, Lactobacillus fructivorans, Lactobacillus paralimentarius, Lactobacillus pentosus, Lactobacillus spicheri, Lactobacillus crispatus, Lactobacillus delbrueckii, Lactobacillus fermentum, Lactobacillus reuteri, Lactobacillus acidophilus, Lc. lactis, Leuc. mesenteroides, Ped. pentosaceus, W. confuse.

Fermented vegetable products

People eat plants, both domesticated and wild, preparing them according to a variety of recipes. Perishable and seasonal leafy vegetables, radish, cucumbers including young edible tender bamboo shoots is conventionally fermented into edible products using the indigenous knowledge of biopreservation. Mostly species of *Lactobacillus* and *Pediococcus*, followed by *Leuconostoc, Weisella, Tetragenococcus*, and *Lactococcus*, isolated from various fermented vegetable food of the world.

Fermented legumes

Among the fermented legumes most of the products are of soybean origin. Two types of fermented soybean foods are produced: soybeans which are naturally fermented by *Bacillus* spp. (mostly *B. subtilis*) with characteristic stickiness, and soybeans which is fermented by filamentous molds (mostly *Aspergillus, Mucor, Rhizopus*).

Bacillus fermented soybean Products are

Products	Place
Natto	Japan,
Chungkokjang	Korea,
Kinema	India, Nepal, and Bhutan
Aakhune, bekang, hawaijar, peruyaan, and tungrymbai	India,
Thua nao	Thailand
Pepok	Myanmar
Mold fermented soybean products are	
Miso and shoyu	Japan,
Tempeh	Indonesia
Douche and sufu	China
Doenjang	Korea
Common non-soybean fermented legume Products in the world are	
Bikalga, dawadawa, iru, soumbala, ugba	Africa
Dhokla, papad, and wari	India
Ontjom	Indonesia
Maseura	India and Nepal

Fermented Meat Products

The major microorganisms involved in meat fermentation are a species of LAB and coagulase-negative cocii, however, yeasts and enterococci are also present in some meat products. Identification is based on the culture-independent approach using the DGGE-method has revealed *Lactobacillus curvatus* and *Lactobacillus sakei* as the main species of LAB involved in the transformation process, accompanied by coagulase-negative cocci *Staphy. xylosus* during meat fermentation and ripening. *Lactobacillus curvatus, Lactobacillus paraplantarum, Lactobacillus plantarum, Lactobacillus sakei, Lactobacillus brevis, Lactobacillus carnis, Lactobacillus casei, Lactobacillus curvatus, Lactobacillus divergens, Lactobacillus sanfransiscensis, Leuconostoc carnosum, Leuconostoc gelidium, Leuconostoc pseudomesenteroides, Leuconostoc citreum, Leuconostoc mesenteroides, Ped. acidilactici, Ped. pentosaceus, W. cibaria, W. viridescens, B. lentus, B. licheniformis, B. mycoides, B. subtilis, B. thuringiensis, E. cecorum, E. durans, E. faecalis, E. faecium, E. hirae* are the dominant LAB in fermented meats and also coagulase-negative staphylococci, micrococci, Enterobacteriaceae in fermented meats. The species of yeasts present in Spanish fermented sausages are *C. intermedia/curvata, C. parapsilosis, C. zeylanoides, Citeromyces matritensis, Trichosporon ovoides*, and *Yarrowia lipolytica.*

Fermented, dried and smoked fish products

Preservation of fish through fermentation, sun drying, smoking and salting is traditionally performed by people living in coastal regions, or near lakes and rivers, and such preserved/fermented fish products are consumed as seasoning, condiments, curry and on the side and sun-dried or smoked fish product such as *gnuchi, sidra, sukuti* of India, Nepal, and Bhutan, *Ent. faecalis, Lactobacillus plantarum, Lactobacillus reuteri, Streptococcus salivarius, Bacillus* species, *Micrococcus, Pediococcus* and yeasts including species of *Candida* and *Saccharomyces* is reported from fermented fish products of Thailand. *Bacillus subtilis, Bacillus pumilus, E. faecalis, E. faecium, Lc. lactis* subsp. *cremoris, Lc. lactis* subsp. *lactis, Lc. plantarum, Lb. amylophilus, Lb. fructosus, Lb. confusus, Lb.corynifomis* subsp. *torquens, Lb. plantarum, Leuc. mesenteroides, P. pentosaceus, Micrococcus*; yeasts—*Candida bombicola, C. chiropterorum,* and *Saccharomycopsis* spp. were isolated from Indian fermented and sun-dried fish products.

Some traditional fermented fish product of the world are

Product	origin
Hentak, ngari, tungtap, bordia, karati, and lashim	India
Jeotgal or jeot or saeu-jeot	Korea,
Plaa-som	Thailand
Shiokara	Japan
Patis	Philippines

Fermented Root Crop and Tuber Products

Cassava (*Manihot esculenta*) root is traditionally fermented staple foods. Cassava root is also traditionally fermented sweet dessert such as *tape* in Indonesia. A mixed culture of *Streptococcus, Rhizopus* and *Saccharomycopsis* produce the aroma in the *tape*, whereas

Sm. fibuligera produce the α-amylase and *Rhizopus* sp. Produce the glucoamylase. *Simal tarul ko jaanr* is a mild-alcoholic food beverage processed from cassava root in Nepal and India. In initial stage fermentation of cassava is dominated by *Corynebacterium manihot. Lactobacillus acidophilus, Lactobacillus casei, Lactobacillus fermentum, Lactobacillus pentosus, Lactobacillus plantarum, Leuconostoc* sp., and *Streptococcus* sp.

Some other fermented products

Vinegar

Vinegar is a popular condiment in the world and it is prepared from any sugar containing substrates and hydrolysed starchy materials by acetic acid fermentation. *Acetobacter aceti* subsp. *aceti, A. oryzae, A. pasteurianus, A. polyxygenes, A. xylinum, A. malorum, A. pomorum* are the dominant bacteria for vinegar fermentation. Yeast species in vinegar fermentation are *Candida lactis-condensi, C. stellata, Hanseniaspora valbyensis, H. osmophila, Saccharomycodes ludwigii, Sac. cerevisiae, Zygosaccharomyces bailii, Z. bisporus, Z. lentus, Z. mellis, Z. pseudorouxii*, and *Z. rouxii*.

Ethnic fermented tea

Tea is the second most popular beverage in the world after water, originated in China and two common species of tea are *Camellia sinensis* var. *sinensis* and *Camellia sinensis* var. *assamica*. However normal black tea is drunk everywhere, but some ethnic Asian communities have special fermented tea such as *miang* of Thailand, *puer* tea and *fuzhuan brick* of China, and *kombucha*. *Aspergillus niger* is the predominant fungus in *puer tea*; *Blastobotrys adeninivorans, A. glaucus*, species of *Penicillium, Rhizopus* and *Saccharomyces*, and the bacterial species *Actinoplanes* and *Streptomyces* is isolated from *puer tea*. *Brettanomyces bruxellensis, Candida stellata, Rhodotorula mucilaginosa, Saccharomyces* spp., *Schizosaccharomyces pombe, Torulaspora delbrueckii, Zygosaccharomyces bailii, Z. bisporus, Z. kombuchaensis*, and *Z. microellipsoides* were isolated from *kombucha* (Kurtzuman et al. 2001, Teoh et al. 2004). The major bacterial genera present in *kombucha* is *Gluconacetobacter* (>85%), *Acetobacter* (<2%), *Lactobacillus* (up to 30%), and yeast populations is found to be dominated by *Zygosaccharomyces* (>95%). *Lactobacillus thailandensis, Lb. camelliae, Lb. plantarum, Lb. pentosus, Lb. vaccinostercus, Lb. pantheris, Lactobacillus fermentum, Lactobacillus suebicus, Ped. siamensis, E. casseliflavus*, and *E. camelliae* are involved in the fermentation of *miang* production. fuzhuan brick tea is fermented by fungi (species Species of Aspergillus, Penicillium and Eurotium)

Cocoa/chocolates

Chocolate is a fermented confectionary product obtained from cocoa beans which require fermentation as one of the first stages. *Lactobacillus fermentum* and *Acetobacter pasteurianus* is the predominating bacterial species during cocoa fermentation. Diverse LAB species appear to be commonly related with the fermentation of cocoa beans in Ghana, and, in fact, that a number of new species have been specify in recent years, for example *Lb. ghanensis, Weissella ghanensis, Lb. cacaonum, Lb. fabifermentans* and *Weissella fabaria. Fructobacillus pseudoficulneus, Lb. plantarum* and *Acetobacter senegalensis* were among the prevailing species at the initial phase of cocoa fermentation and *Tatumella ptyseos* and *Tatumella citrea* were the prevailing enterobacterial species in the beginning of the fermentation. *Bacillus coagulans* are also recovered in vinegar. Yeasts involved during spontaneous cocoa fermentation are *Hanseniaspora uvarum, H. quilliermundii, Issatchenkia orientalis* (*Candida krusei*), *Pichia membranifaciens, Sacch. cerevisiae,* and the *Kluyveromyces* species for flavor development.

Fermented eggs

Pidan is prepared from alkali-treated fresh duck eggs, which have a strong smell of hydrogen sulphide and ammonia.it is consumed by the chinese. The main alkaline chemical reagent used for making *pidan* is sodium hydroxide, which is produced by the reaction of sodium carbonate, water, and calcium oxide of pickle or coating mud. *Bacillus cereus, Bacillus macerans, Staphy. cohnii, Staph. epidermidis, Staph. haemolyticus*, and *Staph. Warneri* are predominant in *pidan*

Alcoholic beverages and drinks

There are 10 major categories of alcoholic beverages consumed/drunk across the world:

1. Nondistilled and unfiltered alcoholic beverages produced by amylolytic starters
2. Nondistilled and filtered alcoholic beverages produced by amylolytic starters
3. Distilled alcoholic beverages produced by amylolytic starters
4. Alcoholic beverages produced by human saliva
5. Alcoholic beverages produced by mono-fermentation
6. Alcoholic beverages produced from honey
7. Alcoholic beverages produced from plants
8. Alcoholic beverages produced by malting (germination)
9. Alcoholic beverages prepared from fruits without distillation
10. Distilled alcoholic beverages prepared from fruits and cereals

Use of starter culture for fermented foods

Starter Culture concept established to preferred over the fermentation process to produce quality products. First Starter Culture is presented in 1893 in Denmark when Emil Christian Hansen working with Carlsberg Brewery isolated the main purified starter culture in the yeast. He recognized and detached yeast responsible for various poor and best quality of beer. "Starter Culture can be characterized as the stain of microorganism chose with stable highlights to deliver alluring characters of food under controlled conditions. starter culture might be isolated from nature/sources tested for desirable character. They were chosen and developed in fermenter. Maximum starter cultures utilized today are developed from lactic acid microbes.

Types of Microorganisms Use in Starter Culture

Dependent on the microorganism employed, starter cultures are

A. Bacteria

B. Yeasts

C. Fungi

A. Bacteria

Different bacterial microbes are utilized as starters in various food products.

Lactobacillus: It is a large group containing gram positive, non-motile, rod-shaped microaerophilic catalase, negative bacteria. First pure culture of lactobacillus is lactobacillus lactis by Joseph Lister from sour milk (Axeleson et al. 1998). It is 2 types of lactobacillus species homofermenters and heterofermenters.

Homofermenters produced lactic acid from glucose, acetic acid and carbon dioxide optimum growth temperature 37°C Ex. lactobacillus bulgaricus, lactobacillus helviticus, lactobacillus acidophilus, lactobacillus thermophilus, lactobacillus plantarum. Lactobacilli is used as a starter in the production of cheese and yogurt.

Heterofermenters produced Volatile substances alcohol and carbon dioxide along with lactic acid. Ex. lactobacillus fermentum, lactobacillus brevis.

They are important as starters in food industry especially in dairy industry because of the following features are: -

(1) Production of lactose from glucose

(2) Production of gas

(3) Resistance to heat even at high temperatures such as production of swiss cheese.

***L.acidophilus*:** It is a non-motile gram-positive rod, homofermenter. Optimum growth temperature 35–38°C, pH 5.5–6. Based on their wide range of pH and temperature they are used in many food productions to obtain taste, flavour, texture etc. It is used in the production of acidophilus milk. The microorganism secretes lactic acid and bacteriocins hence used as a bio preservative in food preparations.

***L.helviticus*:** Homofermenter optimum growth temperature 42–43°C even it can grow at 45°C. It is used in the preparation of Mozzarella and Emmental cheese and yogurt.

***L. lactis*:** It is a gram positive, non-motile, non-spore former, catalase negative, mesophilic and micro aerophilic bacterium. This organism is involved in natural souring of milk even in cultured butter milk.

***Leuconostoc mesenteroides*:** It is a gram-positive rod/coccus based on available nutrients, heterofermenter Facultative anaerobe catalase negative. This organism can tolerate higher concentrations of sugar and salt more than 50 %.

Propionibacterium: They are gram positive rods/cocci, non-motile, non-spore formers, mesophilic and catalase negative. It is used as a starter in swiss cheese i.e. P. freudenreichii subsp. freudenreichii and P. freudenreichii subsp. shermanii for the production of eyes and flavour.

***Streptococci*:** It is spherical, ovoid, gram positive, non motile, non spore former, facultative anaerobe grows at 37°C. The cultures used in foods include Streptococcus lactis and Streptococcus cremoris. These are used in the preparation of cheese and cultured butter milk.

***Streptococcus thermophilus*:** They are spherical in shape occur in pairs or in chains homofermenter facultative anaerobe. Optimum growth temperature is 45°C or more. It is used in the preparation of yogurt.

Staphylococci: It is spherical, ovoid, gram positive bacteria.

B. Yeast

It is a unicellular eukaryotic. It develops in carbon rich sources, for example, glucose, fructose, sucrose and so on. It is obligate aerobe develops in natural or marginally acidic condition. Microorganism gets vitality utilizing carbon sources and discharging carbon dioxide or alcohal. This made this living being as an important device in food industry particularly in the processing of food and dairy items. In the last eighteenth century man knew the production of alcohol utilizing yeast. There is a numerous product prepared by utilizing yeast, for example, Beer, alcohol, wine, bread, kefir and so forth. Selection of Yeast Strain depends upon the type of product and crude material utilized. A good yeast starter ought to have high resilience to, alcohal, sugar, quick carbon dioxide creation and flocculation. Bakery yeast is utilizing the starters of Saccharomyces cerevisiae with the feature of stability, reasonability in dried stored conditions, rapid carbon dioxide production for bread batter raising. Saccharomyces pastorianus is utilized in the processing of Lager brew and Debaryomyces hansenii is utilized in cheddar processing.

C. Molds

They are multicellular filamentous parasites. The body of the parasite comprises of entwined filamentous known as hyphae; all hyphae together known as mycelia. Aspergillus, Penicillium, Rhizopus, yeast (saccharomyces kluveromyces and zygosaccharomyces sps) are the significant contagious starters utilized in food industry. Two types of penicillium are utilized in the processing of cheddar. penicillium camemberti is a white colour present in the outside of Camembert and Brie cheddar. Furthermore, penicillium roquefortii is a blue colour present in the inside of cheddar

Roquefort, Gorgonzola, Stilton. These molds change the surface, smell, taste of cheeses by emitting lipases and proteases. Methyl ketones and free unsaturated fats grant flavour and fragrance to the cheddar. Aspergillus sojae is the starter culture utilized in conventional soy sauce, soy glues and rice wine arrangements. Rhizopus is the starter culture in the processing of Tempeh.

Categorization of starter cultures

Starter culture is classified in differently based on number and temperature of the organism.

Classification of Based on Number

A. Single strain of starter culture.

B. Multi strain of starter culture.

C. Mixed strain of starter culture.

A. Single strain of starter culture

This type of starter culture is comprising of just a single strain. pure starter culture contains at least one strain of same species. Spread milk is utilized for the starter Streptococcus lactis, Cremoris. while Bulgarian buttermilk is prepared with the starter culture of Lactobacillus delbrueckii. Type of starter culture is dependent upon the processing of raw material used to be finished products. Saccharomyces cerevisiae yeast is utilized in the preparation of bread is intended for the production of carbon dioxide or raising the dough and in the production of liquor.

B. Multi strain starter culture.

Multi strain culture contains at least three single strains. for example, these starters are prepared using few strains of various species or various strains of same species. The combination of cultures is use in multi-strain Lactococcus lactis ssp. cremoris and additionally lactis with Leuconostoc cremoris and L. lactis.

C. Mixed strain starter culture

This starter culture contains numerous obscure strains. Proportion of the microorganism additionally change the product to product. Swiss cheese is prepared with mixed culture of Lactobacillus bulgaricus, Streptococcus thermophilus and Propionibacterium shermanii. Kefir is a dairy product. It is prepared with mixed culture contains the Lactic acid microorganisms and Yeast like Lactobacillus bulgaricus, L. helviticus, L. acidophilus, L. kefiranofaciens and Lactococcus lactis sbsp. cremoris and Yeast, for example Kluyveromyces marxinus, Kluyveromyces lactis var. lactis, Debaromyces hansenii edekkera anomala, Saccharomyces cerevisae, Torula spora delbrueckii, Debaromyces accidentalis and so forth.

Classification of based on Temperature

A. Mesophilic starter culture

B. Thermophilic starter culture

A. Mesophilic starter culture

Ideal growth temperature of mesophilic starter culture is 20–30°C. for example Genera of Lactococcus, Leuconostoc and Streptococci. These starters culture is helpful in milk fermentation either exclusively or in blend.

Lactococcus lactis is utilized for cheese production. different types of Mesophilic starters are

(a) **O-type:** These are homofermenters produces lactic acid by and large utilized organism existence is Lactococcus lactis subsp. lactic and Lactococcus lactis subsp. cremoris.

(b) **D-type:** It is a Streptococcus lactis subsp. lactis var diacetyl lactis side by side O-type microscopic organisms. D-type organism produces diacetyl-a characteristic flavour of compound in butter milk. and It is additionally producing carbon dioxide which imparts the mix of sensitive flavour.

(c) **L-type:** It is a Leuconostoc mesenteroides alongside O-type of microscopic organisms and it is produces diacetyl acetic acid, acetaldehyde. and less carbon dioxide than D-type.

(d) **LD-type:** This type of starter culture produces fine mix of fragile flavour and aroma. and the microorganism include Streptococcus lactis subsp. lactis var. diacetyl lactis alongside Leuconostoc mesenteroids. Mesophilic starters are utilized in the production of various cheeses, for example, Brie Camembert, cheddar, Gouda and so on.

B. Thermophilic starter culture

Ideal growth temperature is 40–45°C. These are microaerophilic in nature. and species is Streptococcus and Lactobacillus. Streptococcus thermophilus and Lactobacillus bulgaricus is utilized in the production of Yogurt. and the organism is kept up in 1:1 proportion. lactobacillus bulgaricus develop the flavour part of yogurt and Streptococci produce for the most part acid.

Aspects of starter culture

1. Single microorganism is available subsequently. quality of the product will be uniform.
2. Uniquely structured strains is utilized to produce wanted flavour, texture.
3. Expanded volume of starter culture is decreased the fermentation time.
4. Quality and steadiness of fermentation is kept up.

5. Contamination is disheartened by modifying the parameters like temperature, oxygen, pH, substrate composition and so forth.
6. Greatest substrate is utilized by the organism thus cost of product is reduced.

Function of fermentation in food processing

Fermentation assumes a few significant roles in food handling. The human eating routine can be enhanced through the improvement of flavor and texture in fermented food and preservation of food through the use of lactic acid, alcohol, acetic acid, and alkali-based fermentation processes. food substrates is enhanced protein, essential amino acid, essential unsaturated fat, and vitamins through the biological activity of microbial cultures during fermentation also may increase macronutrients. And, fermentation may detoxify food and reduced cooking times and fuel necessities. In spite of the fact that fermented food is produced worldwide using a wide variety of manufacturing techniques, raw materials, and microorganisms, there are just four main fermentation process normally utilized: - Alcohol, Lactic acid, Acetic acid and Alkali-based fermentation.

Alcohol fermentation It is the production of ethanol, and yeast is the prevalent microorganism utilized (e.g., to deliver wines and lagers).

Lactic acid fermentation (e.g., aged milk and grain products) includes the utilization of lactic acid microscopic bacteria. characteristic lactic acid bacteria utilized in cereal fermentation are: - lactobacillus (rod shaped), streptococcus (sphere in chains), pediococcus (sphere in tetrads), Lactococcus (cocci), leuconostoc (sphere in chains), Bifidobacterium (branched rods), enterococcus (cocci).

A second significant category of bacteria used in food fermentation is acetobacter species. These bacteria produced acetic acid by changing over alcohol to acetic acid within the sight of overabundance oxygen. Alkali fermentation is regularly utilized with fish and seeds, the products of which are consumed as condiments.

In food processing Fermentation plays at least five roles

1. Improvement of the human dietary through improvement of a wide decent variety of flavor, aroma and texture in food.
2. Preservation of generous amount in food through lactic acid, alcohol, acetic acid and alkaline fermentations.
3. Enhancement of food substrates which is commonly present in food. like protein, basic amino acids, essential unsaturated fats and vitamins.
4. Detoxification during food fermentation processing.
5. A diminishing the cooking times and fuel.

Technique innovation in food processing

Fermentation is a deep-rooted nourishment handling innovation, the information on which has been given to progressive ages (Caplice and Fitzgerald 1999; Adebo et al. 2017a). In lieu of the rising population, conventional fermentation procedures are difficult to take into account the necessities of the customer requests, and cutting-edge innovation can guarantee quick conveyance of wanted fermented foods while keeping up consistency in unrivalled quality, tangible traits, and wholesome advantages. Novel food processes is adjusted in food and beverage industries coming the rising of new age technologies and scientific research in food science. Effective utilization of different novel innovations, for example, co-culture, thermophilic fermentation, molecular tools, genetic engineering, mutant selection, and recombinant DNA technologies have made it is possible to design and construct tailor-made starter cultures that perform better than those found naturally. For example, a novel drink was gotten by utilizing blended starter culture, viz., Saccharomyces cerevisiae and Pediococcus acidilactici, S. cerevisiae and Lactobacillus acidophilus, Pediococcus acidilactici and Lactobacillus acidophilus, and S. cerevisiae Pediococcus acidilactici, and L. acidophilus. With the rising competitiveness in the fermentation industry, developments for the preparing innovation for the conveyance of novel, imaginative, high-quality fermented food sources in the world over have taken a significant leap forward.

Thermal and nonthermal applications

The thermal procedures is improving the shelf life of the fermented food, obliterating the pathogenic microorganisms, expanding the metabolic exercises and production of enzymes, just as shortening the fermentation procedure. Few thermal processes include ohmic heating, radio frequency, and microwave heating. However, thermal processing affects the original biochemical, texture, and sensory properties of fermented foods. The best novel nonthermal procedures use in the fermentation industry are high-pressure preparing (HPP), high-pressure carbon dioxide (HPCD), ultrasound, ultraviolet radiation (UV), gamma (γ) irradiation, microwave irradiation, and pulsed electric field (PEF). FDA (Food and Drug Administration, USA) has approved the utilization of γ-irradiation for food handling, it is obtained from cobalt-60 or cesium-137 and UV radiation for surface disinfection. food preservation through ionizing radiation dose up to 7 kilogray is viewed as sheltered by WHO and is accepted by countries, for example, the USA, France, Netherlands, Canada, and so on. Similarly, UV radiation is a nonionizing beam that either eliminates microscopic bacteria or hinders the bacterial multiplication by meddling in the DNA replication through formation of thiamine diamers in the chromosome (US Food and Drug Administration 2007). HPP is use in a range of 100–800 MPa in various temperatures to both solid and liquid food products to ensure sterilization.

It devastates the cytoplasmic layer of the contaminant microorganism. Loss of solute, enzyme inactivation, and protein coagulation are the primary reasons of cell demise. HPCD is viewed as a domain well-disposed and vitality sparing procedure as it utilizes less pressure and CO2 is nontoxic. PEF is pertinent to fluid and semiliquid sample, food material is put between two cathodes, and pulses of high electric field (1–50 kV/cm) for extremely brief periods (2 μs–1 ms) are applied for disinfection (Barbosa-Canovas et al. 2001). Aside from these nonthermal procedures, different natural antimicrobial is added in the form of the extracts from herbs, spices, and vegetables. Extracts of seaweed, capsicum, and green tea inhibit the growth of Salmonella sp. Essential oil of herbs like rosemary, basil, ginger, and sumac contain compounds such as carvacrol, citral, thymol, eugenol, and citric acid well studied for their antimicrobial activity (Gupta and Abu-Ghannam 2012). The nonthermal processing accelerate the chemical reactions such as oxidation, polymerization, condensation, and esterification, augmenting fermentation rates and monitoring the fermentation process, and pasteurization (Oey et al. 2008; Mirza-Aghayan et al. 2015). Alongside the thermal and nonthermal procedures, different methods which are being investigated in the ongoing past incorporate nanotechnology and metabolomics.

Nanotechnology

Nanoscience and nanotechnology have various fields including medicine, cosmetics, agriculture, and food. The scientific noteworthiness of nanotechnology lies in its capacity to study particles at nanoscale (1–100 nm) which are expectedly biologically more dynamic than the bigger measured particles of a similar constituent. although the fact that the fermented food are wealthy in supplements and other dynamic segments, yet these might be lost during processing of the foods. Nanoemulsion arrangements have been utilized effectively to defend nutraceuticals against degradation, providing better stability, and augment the bioavailability and safe delivery of functional ingredients to potential consumers. Another use of nanoscience is nano-sensors which are being utilized for adequately checking and assessing quality parameters of fermented food. These incorporate quantum dots, biosensors for identifying the pathogenic and injurious microorganisms that reason deterioration of food.

Food metabolomics

There have been cognizant and heightened endeavours for advancement and conveyance of functional foods which give extra medical advantages (Adebo et al. 2017a). fermented food is special in their piece with many recognized and characterized components and unidentified ones. The investigation of these components can open up new roads for the food fermented industry. With the presentation of "omics" in food industry, food metabolomics or foodomics as it is called has gotten conceivable to describe the total profile both quantitatively and qualitatively of the food metabolome. With such advanced logical gear, insights concerning the fermented foods like their organization and metabolic cooperation can be recognized that could help in associating with the utilitarian perspectives. Significant components that build up the choice of steps incorporate the sort of study (directed or untargeted), strong or fluid type of test, and the instruments utilized [gas chromatography-mass spectrometry (GC-MS), Liquid chromatography-mass spectrometry (LC-MS), Nuclear magnetic resonance (NMR), etc.]. The procedure of metabolomic investigation incorporates test planning, extraction, information securing, and information examination (Adebo et al. 2017a). Metabolomic study was led to watch the metabolic changes happening during the fermentation of tea (Lee et al. 2011; Tan et al. 2016). Proton NMR, UPLC-Q-TOF-MS, and PCA were applied to think about the fermentation example of tea fermentation, and the blend of investigations uncovered that caffeine epicatechin, epigallocatechin, caffeine, quinate, theanine, and sucrose diminished gallic corrosive and expanded glucose level. Effective fermentation designs and metabolomics have been studied and announced for other conventional fermented foods, for example, kimchi (aged vegetables), koji (aged rice), and doenjang (aged soybean) (Adebo et al. 2017a). An exhaustive comprehension of the procedures and parameters and using this method would help in improving item quality and lessen item costs.

Safety related aspects of fermented foods

Fermented foods for the most part have a generally excellent wellbeing record even of the creating scene where the food sources are fabricated by individuals without preparing in microbiology or chemistry, frequently in unhygienic, defiled conditions. While fermented food are themselves by and large protected, it ought to be noticed that fermented food without anyone else don't take care of the issues of sullied drinking water, conditions vigorously defiled with human waste, ill-advised individual cleanliness in food handlers, flies conveying malady creatures, unfermented food conveying food contamination or human pathogens and unfermented foods, in any event, when cooked whenever took care of or put away inappropriately. Likewise, inappropriately fermented food can be dangerous. In any case, utilization of the rules that lead to the security of aged nourishments could prompt an improvement in the general quality and the healthy benefit of the food supply, decrease of infection and protection from intestinal and different illnesses in Infant.

Assumption of safety aspects of fermented foods

First assumption: -The preservation of food fermentation is identified with few standards. The main act is the food substrates congested with attractive, eatable microorganisms become impervious to intrusion by decay or lethal or food contamination microorganisms. Other, less attractive, potentially sickness creating living beings think that it's hard to contend. A model is Indonesian Tempe the substrate is drenched, dehulled partially cooked soybean. while the initial drenching, the soybeans processing an acid fermentation that decreased the pH to 5.0 or underneath, a pH inhibitory to numerous organisms however profoundly satisfactory to the mold. Following cooking the soy cotyledons are surface dried which hinders development of microscopic organisms that may ruin the item. The fundamental microorganism is Rhizopus oligosporous or related Rhizopus species which weave the cotyledons into a minimal cake that can be cut or cut into solid shapes and used in plans as a protein-rich meaty substitute. These molds can become quickly at moderately high temperatures, for example 40-42°C, unreasonably high for some microbes and molds. The mold uses the accessible oxygen and produces CO, which hinders some potential deterioration microorganism. The mold also delivers amounts of spore dust that penetrate nature and add to vaccination of the ideal microorganisms. The mix of a moderately low pH, no free water and a high temperature of the fermenting bean mass empowers Rhizopus oligosporus to congest the soybeans in 18 h. organism that may some way or another ruin the item can't contend with the mold. The mold additionally creates some anti-toxin action that hinders different organism that may attack and ruin the products. A legitimate concern is mycotoxins that are available in various cereal. Cereal and legume substrates before fermentation. Surely, it is an issue; nonetheless, it isn't the fermenting that produces mycotoxin. It is processed when the legumes grains or cereal are inappropriately reaped or put away. During drenching and cooking of crude substrates before fermentation, numerous potential poisons, for example, trypsin inhibitor, phytate and hemagglutinin are crushed. In this way, fermentation will in general detoxify the substrate and built up a infant formula containing tempe as 39.6% of its weight. Other significant fixings were wheat flour - 31.6%, sugar - 23.3% and vegetable oil - 2.9%. Infants suffering from acute diarrhoea and consuming the formula recovered more quickly than newborn children not getting the tempe formula and furthermore put on weight quickly. Doubtlessly consideration of tempe in infant formulas on a more extensive scale would diminish the general frequency of looseness of the diarrhoea also improve infant/child development of rates and nourishment.

Lactic acid fermentation

Second assumption: Fermentation produced lactic acid are commonly safe. Lactic acid fermentation incorporates those fermentable sugars is changed over to lactic acid by organism, for example *Leuconostoc mesenteroides*, *Lactobacillus brevis*, *Lactobacillus plantarum*, *Pediococcus cerevisiae*, *Streptococcus thermophilus*, *Streptococcus lactis*, *Lactobacillus bulgaricus*, *Lactobacillus acidophilus*, *Lactobacillus citrovorum*, *Bifidobacterium bifidus* and so forth.

This classification is liable for handling and protecting huge amounts of human nourishment and guaranteeing its wellbeing. We are on the whole mindful of the superb wellbeing record of harsh milks/yoghurts, cheeses, pickles and so on. the lactic acid fermentation gives the buyer a wide assortment of flavours, smells and to enhance the human eating regimen. The oldest lactic fermentation is presumably fermented/sour milk. Raw, unpasteurized milk will quickly harsh on the grounds and the lactic acid microscopic organisms present in mature milk sugar, the convert lactose to lactic acid. Within the range of acid and low pH, different microorganism that may attack in the milk and transmit sickness creating life forms are less ready to do as such.

In the event that the whey is permitted to escape or evaporate, the residual curd turns into a crude cheddar. Legumes products and vegetable/fish/shrimp blends are preserved widely by lactic acid fermentation. The great lactic acid/vegetable fermentation is sauerkraut. Crisp cabbage is shredded and 2.25% salt is included. There is an arrangement of lactic acid microscopic organisms that grow. In the first place, Leuconostoc mesenteroides produce lactic acid, acetic acid and CO, which flushes out any lingering oxygen making the fermenting anaerobic. At that point Lactobacillus brevis develop creating progressively acidic. At last *Lactobacillus plunturum* develop increasingly lactic acid and bringing down the pH to underneath 4.0. At this pH and under anaerobic conditions, the cabbage or different vegetables is protected for extensive stretches of time. pickled vegetables, cucumbers, radishes, carrots and practically all vegetables and even some green organic food, for example, olives, papaya and mango are acid fermented within the sight of salt the world over. Indian ranchers can securely safeguard their surplus vegetables by the lactic acid fermentation on the homestead. This improves the stock and accessibility of the vegetable food consistently and improves the sustenance of Indian populace.

Alcoholic fermentations

Assumption 3: Fermentations including generation of ethanol are commonly protected nourishments and refreshments. These incorporate wines, brews, Indonesian tape ketan/tape ketella, Chinese lao-chao, South African kaffir/sorghum lager and Mexican pulque. These are for the most part yeast fermentations yet they additionally include yeast-like molds, for example, Amylomyces rouxii and mold like yeasts, for example, Endomycopsis and now and again microscopic organisms, for example, Zymomonas mobilis. The substrates incorporate diluted nectar, sugar-stick juice, palm sap, organic fruit juices, germinated oat grains or hydrolysed starch, all of which contain fermentable sugars that are quickly changed over to ethanol in natural fermentation by yeasts. About equivalent loads of carbon dioxide and ethanol is delivered and the carbon dioxide flushes out lingering oxygen and keeps up the fermentation anaerobic. The yeasts increase and ferment quickly in different microorganisms, the majority of which is high-impact, can't contend. The ethanol is germicidal along with fermented product stays anaerobic and the product is sensibly steady and protected. With starchy substrates, for example, cereal grains, it is important to change over a portion of the starch to fermentable sugar. This is done in an assortment of ways for example biting the grains to present amylase (Andes locale of South America) where maize is a staple or germination (malting) of barley or the grains themselves in a large portion of the existence where brews are delivered. In Asia, there are in any event two extra methods for fermenting starchy rice to alcoholic beverages. firstly the utilization of mold for example, Amylomyces rouxii which produces amylases changing over starch to sugars and yeast, for example, Endomycopsis jibuliger which changes over the glucose/ maltose to ethanol. The sweet sharp/alcoholic product of rice fermentation is called tape ketan in Indonesia and is expended as a treat. At the point when cassava is utilized as substrate, the product is called tape ketella. This procedure is also be utilized to produce rice wines, Chinese lao-chao and Malaysian tapuyltapai. Another strategy is the Japanese koji process used to ferment rice to rice wine. In this procedure, boiled rice is congested with amylolytic mold Aspergillus oryzae for 3 days at 30°C. The mold secured rice is known as 'koji' is then immunized with a culture of yeast Saccharomyces cerevisiae and water is included. Saccharification of the mold amylases and alcoholic fermentation to the yeast continue at the same time.

Acetic acid/vinegar fermentation

Alcoholic fermentation are not kept anaerobic, microscopic organisms is place with family Acetobacter present in the environment oxidized bits of the ethanol to acetic acid/vinegar. Fermentation along with production of acetic acid yield nourishments or fixing is commonly sheltered. Acetic acid is a bacteriostatic to bactericidal contingent fixation. generally, palm wines and kaffir lagers contain ethanol as well as acetic acid. Acetic acid is a significantly more grounded additive than ethanol. Vinegar is used for pickling and protecting cucumbers and different vegetables. The acetic acid and alcoholic fermentation is utilized to guarantee security in nourishments.

Alkaline fermentations

Assumption 4: Fermentation including exceptionally basic fermentation is commonly sheltered if appropriately fermented. Africa has various significant nourishments/toppings that are utilized to season soups and stews along with fill in as minimal source of protein in the eating routine. Among these are Nigerian dawadawa, Ivory coast soumbara and West African iru made by fermentation of splashed, cooked locust bean Parkia biglobosa seeds with microorganisms having a place with the family Bacillus, commonly Bacillus subtilis. Other alkaline fermentations incorporate Nigerian ogiri made by fermentation of melon seed (Citrullus vulgaris); Nigerian ugba made by fermented oil bean (Pentacletha macrophylla); Sierra Leone ogiri-saro made by fermentation of sesame seed (Sesamum indicum); Nigerian ogiri-igbo made by fermentation of castor bean (Ricinus communis) seeds and Nigerian ogiri-nwan made by fermented of the fluted pumpkin bean (Telfaria occident ale) seeds. Soybeans can be fill in for locust beans. (Steinkraus, 1991). This gathering likewise incorporates Japanese natto, Thai thua-nao and Indian kinema, all dependent on soybean.

The fundamental microorganisms are Bacillus subtills and related bacilli. The creatures are proteolytic and the proteins is hydrolysed to peptides and amino acids. The mixture of high pH and free Ammonia alongside fast growth of the basic microorganisms at generally increase the temperatures above 40°C (review rule 1) make it exceptionally hard for different microorganisms that may ruin the item to develop. Subsequently, the items are very steady and well-protected particularly when dried. These are protected nutriment because that is an unhygienic domain.

Assumption 5: High salt savoury flavoured amino acid/peptide sauces and pastes Expansion of salt (NaCl) in ranges from 13% w/v or higher to protein-rich fermentations brings about a controlled protein hydrolysis that avoids putrefaction, averts advancement of food contaminations, for example, botulism and yields meaty, exquisite, amino acid/peptide sauces and glues that give significant toppings especially to those unfit to manage the cost of a lot of meat in their weight control plans. Substantial/ exquisite seasoned amino acid peptide sauces and glues incorporate Chinese soy sauce/ Japanese shoyul Japanese miso/Indonesian kecap made by fermentation of soybeans or soybean/rice/barley blends with Aspergillus oryzae and fish sauces and glues made by fermentation of little fish and shrimp utilizing mainly the proteolytic gut catalysts of the fish and shrimp. In light of the generally high lysine substance of the soybean, soy sauce/miso is an amazing extra to the nourishing nature of rice. Within the sight of high salt fixation (above 13% w/v yet for the most part in the scope of 20-23%) fish gut catalysts will hydrolyse fish proteins to yield amino acid/peptide, meaty flavoured fish sauces and glues like soy sauces/glues. Halophilic pediococci add to the commonplace alluring fish sauce enhance/smell (Steinkraus, 1996).

Interference for improving the impregnability and nutrient value of fermented foods.

Since fermentation is a multi-step food handling activity, evaluating the security and dietary benefit of fermented nourishments and deciding proper intercessions to improve these ought's to be considered with regards to the whole procedure changing groceries from crude materials to finished results. In the light of present practices for the readiness of fermented weaning producs, three zones of concern can be distinguished: wellbeing, nutrition, and sociocultural components.

The time fundamental for food handling in conventional family fermentation procedures is a requirement on the hours accessible to food preparers who may likewise be occupied with kid care and other family unit exercises. The absence of adequate time can have noteworthy ramifications for the wellbeing and healthful nature of fermented food, as whenever spared by shortening the time of fermentation can endanger the viability of fermentation or the debasement of against dietary components. Some decrease in time consumption may anyway be acceptably accomplished by quickening the fermentation procedure (for example using starter culture) or by an alternate dispersion of work (for example through nourishment creation by a business visionary). Care ought to regardless be taken that no easy routes making the food risky are utilized. Sanitation issues can likewise emerge from preparing or taking care of nourishment under unhygienic conditions. Improvement in food quality and wellbeing can be accomplished by actualizing or advancing standards of good assembling and great sterile practice, and by utilizing the Hazard Analysis Critical Control Point framework (HACCP). The HACCP framework comprises a sanitation confirmation apparatus that can be utilized for the distinguishing proof, appraisal, control of dangers and focuses of the process from crude material to bundling and utilization. The use of HACCP framework to a few fermented African nourishments is completed during the Workshop to characterize the utilization of this way to deal with distinguish the important measures to improve the benefit of fermented food and decide the practices that ought to be the subject of wellbeing training for nourishment handlers. Use of HACCP framework likewise showed the generally high dangers related with the arrangement of certain fermented food and the significance of surveying the safety of every item and every planning distinctly.

From a wholesome perspective, it is probable to improve the supplement thickness of the fermented food while keeping up their alluring semi-fluid consistency by the expansion of amylase-rich flour, gave that the starchy grain or root-crop expected for fermentation is cooked. In like manner, it is probable to improve protein substance of fermented porridges and enhanced legume, for example soybeans or cowpeas.

A significant limitation related the application of fermented food for weaning is that of their poor picture contrasted and business weaning arrangements which, in certain nations, are viewed as prevalent and are related with progressiveness. The instruction of shoppers and nourishment handlers in such manner is basic.

Processing of fermented foods from the ordinary to the industrial level will necessitate several critical steps

1. **Isolation and identification of microorganisms related to the fermentation:** Microorganisms related with indigenous fermentations should be disengaged, appropriately distinguished and preserved ideally in a perceived culture assortment for some time later.
2. **Assurance of the role of microorganisms:** The biochemical roles of microorganisms related with food fermentations should be resolved through chemical examination of items discharged by the microorganisms under controlled research centre conditions.
3. **Determination and hereditary improvement of microorganisms:** Microorganisms liable significant changes in the food during fermentation and allow to chosen and exposed to hereditary improvement designed for amplifying alluring quality traits in the food and the restricting any bothersome characteristics.
4. **Improvement in the process controls of the fermented foods:** Improve the quality and amount of fermented foods can be accomplished by controlling natural factors. for example, temperature, moisture content, air circulation, pH, acidity and impact the movement of microorganism during the fermentation procedure.

5. **Improvement of the nature of crude materials utilized in the production of fermented foods:** Both the quality and amount of fermented foods can be improved by picking crude materials other than those customarily utilized for their generation.
6. **Re-enactment of the research facility in the fermented food:** Before pilot scale generation, and (in a perfect world) after all the five phases above have been all around examined, fermented food might be delivered under research facility conditions. Research centre of fermented foods will include the processing of fermented food source by inoculating microbial detaches having alluring properties, into crude materials.
7. **Pilot arrange creation:** The pilot arrange of the main clear take-off from little scale generation and ought to the founded is the after effect of research facility tests.
8. **Generation or mechanical plant arrange:** The creation arrange is the climax of all the past endeavours and should prompt the accessibility of food of unsurprising and reliable quality on an enormous scale.

Innovations for improvement of health and nutrition in fermented food

Fermented food is utilized as wellspring of sustenance. The interest of fermented food and the recently conceptualized useful food source is in the ascent attributable to its potential medical advantages. These foods have investigated for their different therapeutic properties in the ongoing past. for example, antihypertensive (Ahren et al. 2014), antidiarrheal (Kamiya et al. 2013), blood glucose-bringing down advantages (Oh et al. 2014), and antithrombotic properties (Kamiya et al. 2013), to give some examples. The health advantage of the fermented foods is accounted of bioactive mixes as phytochemicals. for example, phenolics, flavones, unsaturated fats, and saccharides alongside nutrient, mineral, and amino acid in significant amounts when contrasted with their nonfermented structures. (Rodgers 2008; Rodriguez 2009; Sheih 2014; Xu 2015). Fermented food is known to gather bioactive mixes while the handling of the nourishment which is normally missing or present is extremely low sums in natural partners. For example red ginseng roots contain bioactive mixes, like, saponins (ginsenosides) and nonsaponins, although the fermented ginseng roots have indicated expanded degrees of saponins (Oh et al. 2014). These saponins is manage the blood glucose and insulin level. Fermented soybean is a significant staple food expended in Korea, China, Japan, Indonesia, and Vietnam, has been accounted for displaying antidiabetic impacts. Such properties in soybean are somewhat because of both quantitative and qualitatively changes during the handling of the aged item. (Kwon et al. 2010). Dickerson and his partners have shown characterized cell reinforcement and resistant balancing possibilities of fermented papaya (Dickerson et al. 2015). Certain nourishments and refreshments fermented with explicit LAB strains are known to speak to safe guideline and antiallergenic properties (Nonaka et al. 2008). As of late, metabolomics related to microbial environment and genomics are logical instruments proposed toward discovering the ailment battling properties of the bioactive mixes and finding novel ones from fermented foods.

Impact of fermentation on nutritious and minerals

Carbohydrates

The significant of carbohydrate in grains and legumes. the main component of carbohydrate is starch. which gives the most calories in our body. (Chaves-Lopez et al., 2014). Fermentation actuates starch-hydrolysing catalysts, for example, α-amylase and maltase which reduce starch into maltodextrins and simple sugars (Osman, 2011), individually. Studies have demonstrated increase in glucose during beginning stage of fermentation because of starch-hydrolysis and development of enacted maltase and α-amylase. (El-Hag, El-Tinay, and Yousif, 2002; Osman, 2011) The glucose liberated during fermentation is a preferred substrate for microorganisms fermenting the food and could analysed the decreased is carbohydrate after 24 hr of fermentation (Osman, 2011). although, when both glucose and fructose is present during fermentation of pearl millet, microorganisms adopted glucose to fructose as a source of energy considering the level of fructose persist stable. In addition, fermentation decreased starch content in millet varieties with consequent increase in carbon dioxide and ethanol production during fermentation period. Moreover, pH was significantly decreased which activated phytase enzyme. Moreover, Fermentation decreased the starch content of millet while increased the carbon dioxide and ethanol production all through fermentation period. In addition, pH is fundamentally decreased which actuated phytase protein.

Protein

The effect of fermentation in proteins has provide unstable results likely due to different empirical designs, study duration, and variation in the initial protein or amino acid of foods. a lot of studies had reported increase (Chaven & Kadam, 1989; Doudu, Taylor, Belton, & Hamaker, 2003; El-Hag et al., 2002; Pranoto, Anggrahini, & Efendi, 2013), while others recognized decrease (Osman, 2011; Pranoto et al., 2013) in protein and/or some amino acids upon fermentation. It creates the impression that the majority of these impacts may not follow perceptible changes yet, relative changes because of loss of dry material because of microorganism's hydrolysis and utilized the starch and fat as the use of vitality. 24 hr of fermented pearl millet increased the protein content because of loss of sugars (Osman, 2011). and decreased the Lysine, glycine, and arginine (Osman,2011), while methionine was increased after fermentation. While accumulation in protein content may be due to loss of dry material during fermentation and bacterial fermentation is known to increase the lysine content in fermented grains. fermented bacteria produce the lysine content and accumulation its convergence by numerous folds and made grain protein complete. This increase may partly be due to decline the protein complex by microorganism and consequently releasing peptides and amino acids (Pranoto et al., 2013). However, some studies reported that fermented microorganisms is uses the amino acid which is decrease the protein content and quality of some fermented food.

Fermentation increase the digestibility of plant proteins (Ali, El-Tinay, and Abdalla, 2003; Alka, Neelam, and Shruti, 2012; Pranoto et al., 2013). Plant protein has low digestibility comparative with animal protein. low protein digestibility may cause gastrointestinal disorder which may produce the fecal extraction of protein. Henceforth, expanded protein absorbability could lessen the amount of undigested proteins which is conceivably caused food hypersensitivities because of poor retention in gut. (Untersmayr and Jensen-Jarolim, 2008). synthesis of fermentation with other handling procedure has more favourable circumstances. For instance, fermentation prior through cooking is viable in expanding the absorbability of grain protein, bringing it almost to a same amount like meat because of not just demolition of protease (trypsin) inhibitors yet also moderately predigestion of grain proteins by microorganisms during fermentation. (Day and Morawicki, 2018). In that likewise decrease in tannins, oxalate, phytic acid and starch which is complex with proteins and consequently constraining availability by digestive.

larger increase of protein digestibility by fermentation because of average breakdown of complex storage protein into progressively dissolvable structures. because the adequacy of fermentation relies upon actuation in phytase, it isn't surprising that fermented roasted or cooked grains doesn't reduced phytic acid fundamentally as simmering or cooked annihilate phytase. (Egli et al., 2002). Also, the amount of phytic acid reduced upon the onset measure of phytase in the grain. As a grain with low phytase amount, for example corn, rice, oat and millet require either a more protracted fermentation time or the expansion of high-phytase grain to essentially decrease phytates.

Fermentation is possible used of starter culture or normally. natural fermentation is less compelling and nonpredictable however normal type of fermentation is used in develop nations. Pranoto et al. (2013) analysed the effect of Lactobacillus plantarum and natural fermentation for 36 hr on protein digestibility of sorghum flour used is in-vitro models. Protein digestibility is expanded by 92% and 47% utilizing Lactobacillus plantarum and natural fermentation, separately. This expansion is credited to expanded proteolytic catalyst in Lactobacillus plantarum that is reduce tannin. it is complex with protein and separate complex proteins in this manner freeing more peptides and amino acids. Doudu et al. (2003) have recently announced that Lactobacillus plantarum have tannase. it is divide the protein–tannin complex consequently freeing proteins. fermenting micro-flora is use amino acids and protein during fermentation loss of amino acid and proteins.

Minerals

Cereals and legumes are the significant role of minerals in develop nations where these are broadly consuming. Minerals from the plant sources have extremely low bioavailability since they are discovered by complexed with nondigestible material, for example, cell wall of polysaccharides like phytate. Outstandingly, potassium is indispensable piece of phytate where it is covalently reinforced rendering out of by digestive enzymes. The perplexing lattices where these minerals is captured and fortified is generally for the low bioavailability. fermentation is one of the processing techniques that is applied to free these complexed minerals and make them promptly bioavailable.

Fermentation increased the magnesium, iron, calcium, and zinc content in some fermented foods that is regularly consume of India and related with the reduction in the measure of phytates. Notwithstanding, the expansion in mineral substance may be because of loss of dry material during fermentation as organisms disable starches and protein (Day and Morawicki, 2018). Fermentation additionally produce bioavailability of calcium, phosphorous, and iron because of exploitation of oxalates and phytates that complex with minerals and reduced their bioavailability.

There are various mechanisms for the fermentation to builds the mineral bioavailability. Firstly, Fermentation reduce the phytic acid that hinder the minerals free and progressively usable. although, this impact is checked by arrival of tannin during fermentation particularly in high-tannin grains, for example, sorghum (Osman, 2011). The enlargement of tannin during fermentation is ascribed by the hydrolysis of dense tannins, for example proanthocyanidin to phenols. (Emambux and Taylor, 2003). Tannins attach the minerals and decrease the bio-availability build upon the term of fermentation. elongate of fermentation diminished the tannin because activity of microbial phenyl oxidase. However, change of tannins to phenols during the fermentation builds phenol content that cooperates with mineral bioavailability and exploitation of phytates in sorghum with high-tannin content doesn't increment in vitro bio-accessibility of iron (Mohite, Chaudhari, Ingale, and Mahajan, 2013).

Secondly, furthermore fermentation relaxes the intricate framework that implants minerals. Both phytase and α-amylase make the framework free by corrupting phytate and starch, individually. also, some fermented microorganisms is degrading the fiber which slacken the nourishment framework further. (Liang, Han, Nout, and Hamer, 2008). thus, the impact of fermentation relies upon nourishment piece and that other nourishment segments, for example, dietary filaments may hinder the openness of some minerals. To balance these difficulties, germination or incubation of nutrition with polyphenol oxidase (PPO) or phytase during fermentation may help decrease the tannins or phytates, therefore separately, a make mineral bio-accessible. (Towo et al., 2006). although such procedures only may not adequate to lessen the antinutritional factors essentially still it could be essential steps to fermentation. The traditional improvement of sorghum in high tannin isn't adequate to enlarge iron bio-accessibility prior to rather a blend of hereditary alteration and fermentation procedures. (Kruger et al., 2012).

Thirdly, during fermentation require low pH and synthesize iron ingestion because of transformation of ferrous iron, it is less absorbable to ferric iron and it is promptly consumed. as well as fermentation gives ideal pH to enzymatic degradation of phytate. further when fermentation is done before by pounding, mineral bio-availability is also improved. on the grounds that crushing builds grain surface territory and separates cell structure, consequently discharging phytase that degrades phytate (Egli et al., 2003; Hemalatha et al., 2007; Leenhardt et al., 2005; Reale et al., 2007).

Phytochemicals

For a prevalent duration, the significance of phytochemicals (phytonutrients) to the human food and wellbeing is not outstanding. Phytochemicals is a significant plant optional metabolic product is delivered of a phenylpropanoid biosynthesis and shikimate pathways during the development of plants (Zhang, Xu, Gao, Huang, and Yang, 2015). while, During the development, L-phenylalanine, under the impact of phenylalanine ammonia lyase (PAL) catalyzation, change into the cinnamic acid. From that point on, numerous phenolic components, for example caffeic acid, ferulic acid and among others is combined. These can latter changed over into tannins, flavonoids, lignin, and different compounds. Advances is investigate the revealed implication of these phytonutrients to human wellbeing by ideals of their antioxidant properties, (Zhang et al., 2015), cholesterol-bringing down impact (Golzarand, Mirmiran, Bahadoran, Alamdari, and Azizi, 2014; Gunness and Gidley, 2010), and decrease in the production of pro-inflammatory cytokines and immunosuppressive cells (Lesinski et al., 2015). As a result, the intricacy of considering phytochemicals, few reviews have concentrated on contemplating the impacts of conventional handling procedures because of restricted limit in the most lab concentrating these customary strategies. All things considered, fermentation has huge impact on phytochemicals it is both valuable and unfavourable. Fermentation of high-carotenoid bio-fortified maize carry about critical loss of carotenoids (Ortiz et al., 2017) contingent upon the span of the fermentation procedure. bio-fortified fermented maize for 24 and 72 hr held 60%–100% of pro-vitamin A carotenoids. it may, later 120 hr of fermentation, maintenance essentially diminished to somewhere in the range of 27% and 48% relying upon genotypes (Ortiz et al., 2017).

Fermentation for 120 hr generally reduced in-vitro bioavailability of carotenoids in six high-carotenoid biofortified maize genotypes. different instruments have been proposed when clarifying observed low bioavailability of carotenoids from fermented maize. through the interruption of network conceivably by enacted endogenous compounds and microorganisms, it is an expansion in centralization of calcium which may upgrade saponification of free unsaturated fats prompting decreased fat ingestion and increment in unsaturated fat discharge in dung. (Lorenzen et al., 2007). Unsaturated fats is a basic retention of lipophilic carotenoids. a few reports demonstrate that the fermentation enhance the β-carotene ingestion in rodents because of disturbance of nourishment framework. (Phorbee, Olayiwola, and Sanni, 2013). It is along these lines hard to make end dependent on these discoveries since they utilized various models and wellsprings of β-carotene. β-Carotene bioavailability is subject to genotype and preparing technique used to create test nourishment (Ortiz et al., 2017; Phorbee et al., 2013).

The liberating phytochemicals by fermentation (Hubert et al., 2008) is cooperate with proteins, sugars, or minerals making them inaccessible. (Doudu et al., 2003). further, the microorganism fermenting the food it is use these phytochemicals in this manner prompting their decrease. For instance, Hubert et al. (2008) reduction of phytosterols, glycosylated soyasaponins, and tocopherols when soybean germs are fermented for 48 hr utilizing strain of lactic acid microorganisms. in this review, measures of phytosterol is reduce from 4.2 mg/g toward the start of the investigation to 1.1 mg/g toward the finish of the investigation. These creators recommended that the reduced in glycosylated soyasaponins could be because of their transformation from their conjugation, 2,3-dihydroxy-2,5-dihydroxy-6-methyl-4H-pyran-4-one (DDMP) to non-DDMP forms. Soybeans is a rich in isoflavones, for example, genistein, daidzein, and glycitein which are intense antioxidant. Fermentation is accounted for to diminish isoflavones altogether because of hydrolysis of glucosides into aglycone. (Manach et al; 2004).

The impact of fermentation in phytonutrients isn't explicit. the impact of fermentation in antioxidant properties of four cereal utilizing Bacillus subtilis and Lactobacillus plantarum. There is a huge increment in the absolute phenolic acid and all out flavonoid substance with most prominent increment in tests with starter culture. **Dordevic et al. (2010)** exhibited that Lactobacillus rhamnosus is more powerful than Saccharomyces cerevisiae. it is discharge absolute phenolic acid during the fermentation of cereals. throughout the fermentation, microorganisms separate cereal grain and lattices prompting arrival of bound phytochemicals (Dordevic et al., 2010). Lactobacillus plantarum and Bacillus subtilis has been recently reported to have β-glucosidase it is divided into glucoside bonds among phytochemicals and sugars. it is a manner of discharging phytochemicals. Along these lines, the sufficiency of fermentation is expand when antioxidant properties of nourishments is investigated as a financially savvy approach to lessen oxidative stress inside in the body subsequent to expending such nourishments.

Benefits of fermented food

- **Preserves food:** Fermentation expands the timeframe of realistic usability of food which save the utilizing of lactic acid, alcohol and acetic acid.
- **Addition microbes to the gut:** People get great microscopic organisms to balance the digestive system and appropriate absorption.
- **Increases micronutrients:** Lactic acid microscopic organisms can expand levels of nutrients in food, particularly B nutrients.
- **Makes food increasingly digestible:** Food contains numerous antinutritive elements that do not digest by the people so these Lactic acid bacteria helpful to them. As the lactose is present in milk and it is separated into two simple sugars – glucose and galactose, a few people having issues of lactose intolerance they can without much of a stretch expend fermented milk item. Additionally, numerous organisms produce chemicals that separate cellulose in plant food, which people can't digest, into sugars
- **Development of taste:** It can make food agreeably harsh or tart, and creates flavour.
- **Eliminates Anti-supplements:** Anti-supplements are the natural or synthetic compounds that meddle with the retention of supplements, which can be decimated by fermentation. For instance, Phytic acid, which is found in legumes and seeds, it ties minerals, for example, iron and zinc, diminishing their assimilation when eaten. In any case, phytic acid can be separated during fermentation, so the minerals become accessible. Diminishes cooking time: Foods that are intense, hardened to digest or unpalatable crude are improved by fermentation, and lessening the requirement for cooking.
- **Produces carbon dioxide:** Fermentive yeast and heterofermentative microbes produce carbon dioxide. Carbon dioxide can be utilized for raising bread and carbonating drinks.

Raw Material	Stability	Safety	Nutritive value	Acceptability
Meats	++	+	-	(+)
Fish	++	+	-	(+)
Milks	++	+	(+)	(+)
Vegetables	+	(+)	-	(+)
Fruits	+	-	-	++
Legumes	-	(+)	(+)	+
Cereals		-	(+)	+

++ definite improvement; +- Usually some improvement; (+) some case of improvement; - no improvement

Fermented food on cereal

Cereal is an important crop of human food throughout the world. Grains becomes 73% of the complete world collected zone and contribute over 60% to the world food production giving dietary fiber, protein, vitality, mineral and nutrient required for human being. storage grains are metabolically inactive with extremely low water activity and moisture content of around 9 -12%. low water activity and moisture content anticipates the development of organisms and the proteins are additionally inactivate at this stage. This is the resting stage of the grain. Increase the water content of the grains causes them to absorb water which eventually invigorates the enzymes enthusiastically and development of microorganisms. This process empowers fermentation to begin and followed by size reduction through milling and sometimes the use of specific organisms and enzyme actions. The nutritional properties of cereals and the sensory properties of the products are in some cases substandard or poor in correlation with milk and dairy products. The explanations for this are the lower protein content, insufficiency of certain basic amino acids, the nearness of decided antinutrients (phytic acid, tanins and polyphenols) and the coarse nature (Blandino et al; 2003).

Fermentation can effectsly affect the dietary benefit of nourishment. Microbial fermentation helps a reduction in the amount of carbohydrates as well as some non-digestible poly- and oligosaccharides. The latter decreases symptoms are, for example, stomach expansion and flatulence. Certain amino acids might be combined and the accessibility vitamin B might be improved. Fermented grain by lactic acid microorganisms has been held for to build free amino acids and by proteolysis and additionally for metabolic synthesis. The microbial mass can also supply low molecular mass nitrogenous metabolites by cell lysis (Mugula et al 2003). Fermentation has been improving the nutritional benefit of grains, for example, wheat and rice, fundamentally by expanding the substance of the basic amino acids lysine, methionine and tryptophan. Fermented rice by lactic acid microbes upgrades the flavour, nutritive worth and accessible lysine content.

Common fermentation of maize expanded all out dissolvable solids and non-protein nitrogen and marginally expanded protein content. 16 hr fermented batter had more significant levels of the egg whites in addition to globulin division, demonstrating that common fermentation of maize brings about improved in the nutritional quality of the grain. Also, maize protein edibility is raised. (Yousif et al 2000).

Fermentation additionally gives ideal pH conditions to enzymatic debasement of phytate which are present in the grains as buildings with polyvalent cations, for example, iron, zinc, calcium, magnesium and protein. similarly, reduction in phytate and may build the measure of soluble iron, zinc and calcium of considerable amount. During the fermentation, decrease of phytic acid content may somewhat be expected to phytase action as it is recognized to be controlled by a wide scope of microflora.

Ideal temperature for phytase movement has been known to run between 35 °C and 45 °C (Sindhu, et al 2001).

During the lactic acid fermentation, the tannin levels might be decreased, and prompting expanded assimilation of iron, aside from in some high tannin cereals, where no improvement in iron accessibility has been observed (Nout et al; 1997). Lessening impact of fermenting on polyphenols can be due to the activity of polyphenol oxidase present on the food grain or microflora (Sindhu, et al 2001).

Fermentation of Conventional foods are arranged the normal sorts of grains, (for example, rice, wheat, corn or sorghum) these are outstanding in numerous portions of the world. Some are used as colorants, flavours, drinks and breakfast or light dinner nourishments, while couple of them is utilized as basic nourishments in the diet. (Blandino et al; 2003).

Fermentation of cereals by lactic acid is a considering of long-time handling technique it is being used in Asia and Africa for the generation of nourishments in different structures, for example, refreshments, gruels and porridge. Cereal grains, basically maize, sorghum, or millet grains are absorbed clean water for 0.5–2 days. Soaking soften the grains and makes them simpler to crash or wet- mill into slurry from which hulls, bran particles, and germs can be expelled by sieving systems. During the slurry or dough making stage goes have been 1–3 day, at this time blended fermentation together with lactic acid fermentation happens.During fermentation, the pH diminishes with a synchronous increment in causticity, as lactic and other natural acids amass because of microbial activity.

The Conventional foods made from cereal grains commonly lack flavour and fragrance. fermented cereal contains a few unstable compounds are formed, which add to a complex mixture of flavours in products. The nearness of aroma represented to a diacetyl, acetic acid and butyric acid makes fermented cereal-based products progressively appetizing ((Blandino et al; 2003). The proteolytic action of fermentation microorganisms frequently in blend with malt catalysts may create antecedents of flavour compounds, for example, amino acids, which might be deaminated or decarboxylated to aldehydes and these might be oxidized to acids or diminished to alcohols [Mugula et al 2003]. And biochemical pathways prompting flavour production can be help in making the appropriate determination of starter culture. however, the finished product of lactic acid fermentation depends on the chemical arrangement of the substrate (carbohydrate content, present of electron acceptors, nitrogen receptiveness) and the natural conditions (pH, temperature, aerobiosis/anaerobiosis), controlling of which would enable explicit fermentation to be directed towards a progressively alluring product.

Compounds formed during cereal fermentation

Organic acids	Alcohols	Aldehydes and ketones	Carbonyl compounds
• Butyric	• Ethanol	• Acetaldehyde	• Furfural
• Succinic	• n-propanol	• Acetaldehyde	• Methional
• Formic	• Isobutanol	• Formaldehyde	• Glyoxal
• Valeric	• Amyl alcohol	• Isovaleraldehyde	• 3-Methyl butanal
• Caproic	• Isoamyl alcohol	• *n*-Valderaldehyde	• 2-Methyl butanal
• Lactic	• 2,3-Butandieol	• 2-Methyl butanol	• Hydroxymethyl furfural
• Acetic	• β-Phenylethyl alcohol	• *n*-Hexaldehyde	
• Capric		• Acetone	
• Pyruvic		• Propionaldehyde	
• Palmitic		• Isobutyraldehyde	
• Crotonic		• Methyl ethyl ketone	
• Itaconic		• Butanone	
• Lauric		• Diacetyl	
• Heptanoic		• Acetoin	
• Isovaleric			
• Propionic			
• n-Butyric			
• Isoburyric			
• Caprylic			
• Isocaprilic			
• Pleagronic			
• Mevulinic			
• Myristic			
• Hydrocinnamic			
• Benzylic			

Microorganisms for cereal fermentations

The basic fermentation procedure includes the enzymatic activities of lactobacilli, leuconostoc, pediococci, yeasts and molds. Their metabolic activities bring about the formation of short chain unsaturated fats, for ex. lactic, acidic, butyric, formic and propionic acid. The pH of these food is decline to consideration of 4 or less. Acids formed during the fermentation process bring down the pH hence reduced the development of deterioration organisms. The pH requirements of a number of pathogens are shown in Table 4 Lactic acid microscopic organisms are industrially significant microorganisms that are utilized everywhere throughout the world in a huge assortment of industrial food fermentations (Hugenholtz et al 2002). The essential activity of the culture in food fermentation is to change over carbohydrates to wanted metabolites, for example acetic acid, lactic acid or CO2. The culture used in the food fermentation is additionally contributing by auxiliary respond to the arrangement of flavour and texture. Lactic acid microbes are considered to an advantageous physiological impact, for example antimicrobial activity, improving of immune strength and counteractive action of cancer and lower serum cholesterol levels (Kaur et al 2002).

The proposed wellbeing and nutritional advantages of Lactobacillus species are

- enzyme (lactase) arrangement,
- colonization and support of the reasonable intestinal microflora,
- competitive avoidance of unfortunate microorganisms,
- microbial impedance and anti-microbial activities,
- pathogen free,
- immuno-incitement and balance,
- cholesterol decrease/expulsion.

Lactobacillus spp. assume a noteworthy role in the vast majority of the fermented cereal. The *Lactobacillus bulgaricus* and *Streptococcus thermophilus* are appropriate microscopic organisms for rice fermentation on account of their absence of amylase, which is essential for saccharification of rice starch. It was indicated that Lactobacillus plantarum and Pediococcus spp. overwhelm the last phases of maize dough fermentation may along these lines be responsible for the quick acidification of inoculated dough (Nche et al 1994). Mixed culture of lactic acid microorganisms and propionic microbes seemed, by all accounts, to be a potential option alongside traditional starters in rye sour dough fermentation, accommodating an improved mould-free shelf-life of bread (Javanainen et al 1993).

In case when the cereal grain is utilized as common mechanism for lactic acid fermentation, amylase should be included previously or during fermentation or amylolytic bifidobacteria should be utilized because these microorganisms contain enough amylase which is important for saccharification of the grain starch (Kim et al 2000). The utilization of amylolytic lactic acid microscopic organisms as a starter culture offers another option by consolidating both amylase production and fermentation of microorganism (Santoyo et al 2003).

Fermentation of cereals with unadulterated yeast cultures has been appeared to build the protein substance of the fermented products. The conceivable capacity of yeasts in fermented foods and drinks are (Jespersen et al 2003):

- fermentation of carbohydrates (arrangement of alcohols and so on.),
- production of aroma compounds (esters, alcohols, natural acids, carbonyls, and so on.),
- stimulation of lactic acid microscopic organisms giving fundamental metabolites,
- inhibition of mycotoxin producing moulds (supplement rivalry, poisonous compounds, and so on.),
- degradation of mycotoxins,

- degradation of cyanogenic glucosides (linamarase action),
- production of tissue-debasing chemicals (cellulases and pectinases) and
- probiotic properties.

It has been recommended that the multiplication of yeasts in food is supported by the acidic condition made by lactic acid microbes while the development of microscopic organisms is invigorated by the presence of yeasts, which may give growth factors, for example, nutrients and soluble nitrogen compounds. Association of lactic acid microbes and yeasts during fermentation may also contribute metabolites, which could give taste and flavour to fermented food.

pH tolerance of some microorganisms

Organism	Minimum pH	Maximum pH
Escherichia coli	4.4	9.0
Salmonella typhi	4.5	8.0
Campylobacter jejuni	2.3	
Shigella sp.	4.5	8.0
Streptococcus lactis	4.3-4.8	
Lactobacillus sp.	3.0	7.2
Yeasts	1.5	8.0–8.05
Moulds	1.5-2.0	11.0

Cereal Based Fermented Foods and Beverages

Products	Microorganisms	Substrates	Regions
Ogi	*Lactobacillus plantarum, Lactobacillus fermentum, Lactobacillus* spp., *Saccharomyces cerevisiae, Candida mycoderma*	Maize, Sorghum/ Millet	West Africa
Enjara	*Candida guillienmandi*	Sorghum	Ethiopia
Pito	*Geotrichum candidum, Lactobacillus, Candida*	Sorghum/Millet	Nigeria, Ghana
Fura	*Lactobacillus plantarum, Pedicoccus, Leuconostoc, Streptococcus, Enterococcus, Pichia anomala,*	Maize/Sorghum	Nigeria
Sake	*Saccharomyces* spp.	Rice	Japan
Dosa	*Leuconostoc mesenteroides, Streptococcus faecalis, Torulopsis candida, T. pullulans*	Rice and Bengal gram	India

Products	Microorganisms	Substrates	Regions
Idli	*Leuconostoc mesenteroides, Streptococcus faecalis, Torulopsis, Candida, Tricholsporon pullulans*	Rice grits and Black gram	South India, Sri Lanka
Miso	*Aspergillus oryzae, Torulopsis etchellsii, Lactobacillus* spp.	Rice and Soya	Japan, China
Tarhana	*Zygosaccharomyces rouxi Streptococcus thermophilus Lactobacillus bulgaricus*	Wheat flour	Turkey
Soya sauce	*Aspergillus oryzae*	Soybean, Wheat	Japan, China & India
Masa Agria	*Lactobacillus spp., L. Plantarum, L. Fermentum*	Maize, Rye	Africa
Bread	*Saccharomyces cerevisiae*	Barley, Wheat	Middle East, North Africa, Europe, America, Australia, Southren, Africa
Togwa	*Lactobacillus fermentum, Pediococcus pentosaceus*	Sorghum, Millet, Maize	Tanzania
Yos a	*Bifidobacteria Leuconostoc mesenteroides,*	Oat bran	Finland
Dhokla	*Streptococcus faecalis, Torulopsis candida, T. pullulans*	Bengal gram and Rice	South India, Srilanka
Beer	*Saccharomyces spp.*	Cereals grains	All over world
Ambali	*L. mesenteroides; L. fermentum and Streptococcus faecalis*	Ragi flour and rice	India
Nan	Yeast (Saccharomyces cerevisiae) and LAB	Wheat flour,	North India.
Tao tjo	*Aspergillus oryzae.*	Roasted wheat meal or glutinous rice and soybean	East India.

Nutritional properties change during fermentation of cereal

Change in pH, titratable acidity and dry matter

After fermentation of cereals or cereal-legume mixes few investigators have studied the adjustments in pH, titratable acidity, and dry issue yield. During fermentation pH decreases with attendant increment of acidity as the lactic acid aggregates because of microbial activity. The respiratory and physiological activities of fermenting organisms use some portion of the food supplements, causing a decline in dry matter yields. During natural fermentation, food contamination flora and coliforms additionally grow with the lactics. These microorganisms should be wiped out to make fermented foods safe for utilization. This can be accomplished by fermenting the suppers for a specific period to arrive at a pH and additionally acridity which is inhibitory to these organisms. An inhibitory pH range of food contamination microscopic organisms is viewed as 3.6 to 4.1. The period required to arrive at this pH range is governed by nature of fermenting material and its microflora, solids water content and temperature of fermentation. These factors additionally impact the yields of dry matter. Overall, the dry matter yields should be higher to maintain a strategic distance from food losses. consequently, fermentation conditions can be deliberately standardized for every grain and cereal-legume mixes as for the solids water proportion and the time and temperature of fermentation. In any case, the extent of water in relation to solids in fermenting medium essentially impacted the degree of acidity and dried matter yields. With less water, more elevated levels of acids were available, repressing the flora.

Fermentation change in protein content

Proteins is comprised of amino acids. The quality of food proteins is estimated by amino acid and digestibility. Nutritionally, proteins are separated into complete, partially complete, and deficient proteins. Complete proteins give all the basic amino acids to support growth and maintenance, though mostly complete proteins can only support the latter. Improved digestibility of proteins adds to a higher protein quality in fermented foods, and it is supported by numerous individuals invitro and invivo observations (Divyashree et al., 2013).

By and large, wheat is inadequate in lysine content, yet fermentation can be improve the lysine content in wheat and increased the protein efficiency ratio (PER). PER is increased the protein quality, and fermented maize have higher protein quality in comparative of nonfermented maize. fermented mixed culture of pearl millet flour with Saccharomyces diastaticus, Saccharomyces cerevisiae, Lactobacillus brevis, and Lactobacillus fermentum have to improve or biologic utilization in rats. However, there are a some reports that suggest fermented foods might not have any specific advantage over nonfermented foods in terms of protein quality Guermani et al. (1992) didn't discover any distinction in PER in rats fed fermented or nonfermented "Okara" samples (okara is the fibrous residue obtained as a by-product during the preparation of milk from the soybeans).

Change in cereal carbohydrates during fermentation

Since the microorganisms is responsible for the degradation of fermentation and the use of starch as a source of energy, there is a decrease in the total starch content of fermented grains. fermentation of single and mixture culture of pearl millet flour with yeast and Lactobacilli spp. have been shown to increase total soluble sugars significantly, reducing and nonreducing sugar content, with a simultaneous reduction in starch. Certain ethnic populaces in Africa use porridge produced using grain or a starchy tuber as a staple food. For adult, the grain flour is cooked as a thick porridge, which is diminished to fluid gruel for children to facilitate feeding. The energy density of such a porridge or gruel relies upon the flour and water proportion and can be low, particularly for thin gruels, compromising dietary intake. Human energy requirements per kilogram in body weight is most prominent during the growth development (about 100 kcal/kg body weight/day for infants in correlation with 40–45 kcal/kg body weight/day for adults), implying that young children need calorie-thick food to meet their fundamental vitality needs and beat their restricted stomach capacity. some fermented food has a much lower consistency than nonfermented starches and subsequently offer the benefit of a healthfully thick food, which can be effectively processed by children (Svanberg and Lorri, 1997). The expansion of more flour to gruels while keeping a comparative consistency would likewise guarantee intake of different supplements innate in the grain, for example, proteins, nutrients, and minerals. This idea has been utilized with amylase-rich flour (ARF) got with germinated grains. A modest quantity of ARF added to weaning food lessens the consistency of the gruel, encouraging the option of more flour and expanding caloric thickness. A mix of both, ARF and fermentation can improve the dietary nature of advantageous food enormously, which is of most extreme significance for some, oppressed populaces subsisting just on grains and tuber crops for the greater part of their suppers.

Cereal contain varying amounts of indigestible oligosaccharides, for example, stachyose, verbascose, and raffinose, which cause gas, diarrhea and digestive problems related issues. In this way, fermentation partially digests carbohydrates, lessens thickness, and increments caloric thickness of food.

Changes in cereal vitamins during fermentation

Fermentation can likewise realize changes in vitamin content, in spite of the fact that the degree of increment or abatement differs relying upon factors such as food material, starter culture, duration and temperature of fermentation. Differing results have been found regarding changes in the vitamin contents of the fermented products. During fermentation, certain microorganisms produce higher rate of vitamins than others. A lessening is the thiamine substance of dough (Indian flat bread made with whole wheat flour) fermented utilizing yeast or undergoing self-fermentation, though an expansion in thiamine and riboflavin was accounted for following fermentation of sorghum, pearl and finger millets. The thiamine and niacin level of the fermented rice were lower than in unfermented controls. Murdock and Fields (1984) reported a 50% increment in thiamine and riboflavin in dhokla and ambali after fermentation, while Gupta, Rudramma, Rati, and Joseph (1998) revealed a reduced. Content of vitamin B12, riboflavin, and folacin expanded in lactic acid fermented maize flour, yet levels of pyridoxine were diminished. Kefir made from 10 different kefir grain cultures indicated a significant increment (>20%) in pyridoxine, cobalamin, folates, and biotin and a decrease surpassing 20% for thiamine, riboflavin, and nicotinic and pantothenic acids, depending on the culture utilized. There is a 40% expansion of the thiamine with two way of the cultures, yet riboflavin indicated just a little increment with these two cultures, while pyridoxine expanded over 120% with three distinct cultures (Kneifel et al., 1991). In this manner, fermentation can expand the substance of a portion of the B-complex vitamins, however may reduce different vitamin or have no impact on their accessibility.

Changes in cereal minerals during fermentation

Mineral substance isn't influenced by fermentation except if salts is added to the product during processing or draining, when the liquid portion is isolated from the fermented food. when fermentation is proceed in metal compartments, minerals are solubilized by the fermented product, which may cause an obvious increment in mineral substance. The dissolvability of metals increases in consider to the acidic nature of the fermenting material. Nonetheless, fermentation affect the bio-accessibility of minerals because of a elimination in anti-nutritional factors. There is frequently a decreased the phytates, polyphenols, tannins and, most likely, oxalates, decreasing the binding of minerals and enabling more minerals to the absorbed by the small digestive organs. This is of incredible practical esteem in the determination of achievable processing to improve assimilation of iron, zinc, and minerals in the diet. Vaishali, Medha, and Shashi (1997), who examined the benifit of natural fermentation on invitro zinc bioavailability in cereal–legume mixtures, found that fermentation increased zinc solubility (2–28%) and take-up altogether (1–16%). Divyashree et al. (2013) additionally observed an expansion invitro in bioaccessible of iron and calcium from fermented chapati dough. Hemalatha, Platel, and Srinivasan (2007) fermented iddli and dosa contains (50–71%) zinc and (127–277%) iron, Indian fermented products based in rice and black gram dhal in comparison with their unfermented partners. Aside from a decreased in antinutrients, the acidic nature of fermented foods expands the mineral bio-accessibility. Consequently, by and large, fermented foods impact mineral take-up by increased bioavailability, in spite of the fact that the actual amounts not increment.

Changes in cereal antinutritional factors during fermentation

Despite the fact that cereals and legumes have high amount of minerals, but their bioavailability is low because of the phytates, polyphenols, tannins, and dietary fiber. These bind minerals, lessening their absorption in the gastrointestinal tract. During fermentation, phytases hydrolyze phytates, favouring their absorption. Diminished pH during fermentation likewise aids the debasement of phytates. The procedures associated with fermentation, for example, soaking, grinding, and heat treatments (cooking) additionally help bring down the antinutritional substance and increment minerals bioavailability. A blend of cooking and fermentation improves the nutrient quality of sorghum and decreases the substance of antinutritional components to a sheltered level contrasted and different techniques for processing. The fermentation of bread dough decreases phytic acid, reduce the ability of phytic acid depend on the yeast and fermentation time (Harland and Harland, 1980). Reported decreases in phytates 28% to 50% whole wheat flour (Divyashree et al., 2013). On the other hand fermentation by S. diastaticus pursued and L. brevis eliminates phytic acid from pearl millet flour following an expansion in fermentation time.

Aside from bringing down mineral retention, phytates is reduced the activity of digestive enzymes, for example, trypsin, alpha-amylase, and beta-galactosidase because of the formation of complex. Tannins and polyphenols also restrain catalysts by framing complex with these proteins, bringing about their deactivation, reduced protein solubility and digestibility, and absorbable ions (Dlamini, Taylor, and Rooney, 2007; Dykes and Rooney, 2006). The enzymes repressed by tannins and polyphenols incorporate pepsin, trypsin, chymotrypsin, lipases, glucosidase, and amylase. Generally, fermentation improves the wholesome nature of food by corrupting antinutritional factors.

Functional properties change during fermentation of fereals

The functional properties like, nitrogen solubility, absorption and retention of water, oil absorption, bulk density, emulsifying capacity and emulsion stability, foaming capacity and stability, viscosity, and so forth of the constituents use to developed newer processed foods are significant. The preparing of crude materials, for example, grains and their suppers by dehulling, processing, cooking, confining significant seed segments, like protein disengages and focuses, high protein flour part through air order, germination, and maturation regularly lead to adjustments of effective properties. The constitution of functional properties of any crude or handled extender would decide its application in a specific food framework.

The legumes or oil seed meals are frequently blended in with cereal to redesign the dietary nature of conventional food. These extenders need to have the alluring properties to make them compatible in food formulations. The grains or cereal-legume mixes are frequently exposed to characteristic lactic acid or yeast fermentation to set up an assortment of conventional food. Furthermore, fermented and dried food are used as enhancements or extenders of the arrangement and assortment of processed foods, for example, weaning food sources, bakery items or breakfast food sources. Henceforth, investigations on the changes in the functional properties of suppers of grain and cereal-legume mixes due to fermentation are significant.

Starch and proteins are the significant component of cereals, legumes, and defatted oil seed meals governing their functional properties. (Fleming et al.1974) revealed that the water absorption limit is ascribed to the protein of the food material. (Khan et al.1975) found that the water absorption capacity of wheat flour expanded with expansion of defatted groundnut flour. (Chandrasekhar and Desikachar1983) detailed that the water absorption capacity of sorghum meal is mostly administered by its starch content. The carbohydrate fraction has been embroiled in affecting the water retention capacity of food material. (Kilara, et at 1972) The high protein content flour is support protein-lipid interactions and add to the expanded fat absorption ability to the food material. The amount of food would be impacted of the number and bulk density of protein bodies and starch granules. Also, the foaming stability as well as emulsifying properties is known to be impacted of starch and soluble proteins and other seed segments. The delicate supple surface recognized is the conventional product idli prepared to the fermented rice in addition to black gram mixture has been attributed to the stabilization of the foam network stability to the surface-active globulins and the carbohydrate fraction called arabinogalactan in black gram. The proteins and carbohydrates undergo significant hydrolytic changes during natural fermentations, the desirable or undesirable modifications in the functional properties of meals are very self-evident.

A harmony between surface activity and consistency is essential to acquire idli with low bulk density and supple surface. processing of germination and fermentation may reduce the surface activity of protein and consistency of arabinogalactan. The overnight fermentation didn't antagonistically influence the functional properties of surface-active protein and arabinogalactan of the black gram. (Sarasa and Nath 1985) be that as it may, observed a reduction in consistency during a 20-h fermentation of mixes of rice or basic millet with black gram and French bean suppers. These reports show that the fermentation period is significant in keeping up the attractive functional properties of idli batter.

(Ahmed and Ramanathan 1988) contemplated the distinctive functional properties of naturally fermented sorghum meal and composite flour made up of 30% eatable defatted groundnut flour and 70% sorghum flour. The fermentation of sorghum flour reduces its ability to water retention ability, bulk density, and both its capacity to form foam and its stability. However, fat ingestion of fermented sorghum flour is increased. Fermentation expanded the nitrogen solubility of the composite flour in the alkaline pH region. The expanded protein of the composite flour improved the functionality of the meal or its better water-absorption capacity, water-retention capacity, and its fat-absorption capacity compared with sorghum flour. other hand fermented composite flour has increased the water absorption ability but decreased the water retention capacity and fat absorption ability. The fermented flour exhibit better foaming properties, however didn't improve its emulsion stability.

Health benefit of cereal based fermented foods

As a probiotic

Lactobacillus plantarum has been connected to diminished inflammatory bowel, little gut bacterial abundance in children, decreased issues for sufferers of touchy gut disorder, and positively affected the safe frameworks of those experiencing HIV. Bifidobacteria has been connected to diminish cases neonatal necrotizing enterocolitis. Numerous strains of probiotics have been legitimately connected to diminished episodes of stomach related protests including loose bowels. Lactobacillus acidophilus additionally demonstrated a safeguard impact for polyps, adenomas, and colon disease.

Wellspring of nutrition

Fermentation increment the digestibility and dietary benefit of grains-based food. Grains based fermented foods secure the body against age-related illnesses, for example, diabetes and cardiovascular sicknesses. Cereals also contain micronutrients, for example, nutrient vitamin E, folates, phenolic acids, zinc, iron, selenium, copper, manganese, carotenoids, betaine, choline, sulfur amino acids, phytic acid, lignins, lignans, and alkyl resorcinol which give different health benefits.

Flatulence lessening impact

Fermented beans are used for the preparation of Tempe, during this process trypsin inhibitors is inactivated and the measure of a few oligosaccharides which generally cause flatulence are fundamentally diminished.

Anti-cholesterolemic impact

Hepner et al. (1979) announced hypercholesteraemic impact of yogurt in human subjects getting a multi element dietary supplement. Concentration of the supplementation of infant formula with Lactobacillus acidophilus, that the serum cholesterol in infant formula was diminished from the 147 mg/ml to 119 mg/100 ml.

Anticancerogenic property

Anticarcinogenic property of fermented foods demonstrating potential role of lactobacilli in diminishing or taking out procarcinogens and cancer-causing agents in the alimentary canal.

Provide functional components

Fermented cereals can likewise contain a high mineral substance and for the most part have a lower fat content than their dairy-based partners, yet grains are commonly ailing in basic amino acids. These types of refreshments can also normally give plant-based functional components, for example, fibre, nutrients, minerals, flavonoids and phenolic mixes, which can impact oxidative pressure, inflammation, hyperglycaemia and carcinogenesis.

Future aspect of cereal based fermented food

In the developed countries, by means of huge weight issue and furthermore for keeping up ordinary and sound wellbeing, various formulation and exercises are coming up, uncommonly conveying dissolvable fibres to the shoppers by means of various nourishments like grains and cereal items containing antioxidant. Cereal like wheat, maize, rice, oats and so forth are currently utilized in arrangement of nourishment that are comparative in appearance to traditional nourishment and utilized in typical eating regimen however have an additional bit of leeway of helping physiological capacities alongside giving sustenance. Dietary patterns can radically lessen medicinal services uses if people somehow managed to change their control diets plans dependent on a current information on nourishment. In this day the advancement and usage of various cereal based useful nourishments is a difficult assignment. Innovation of more up to date advances for processing of cereal to improve their dietary benefit versus their adequacy by the end users will be the center territory soon.

Fermentation of pulses

Pulses are consumable seeds of the family Leguminosae (Fabaceae) that incorporate significant grain species that rank second after cereal in their significance for human nutrition. As reported by Food and Agricultural Organization of the United Nations, pulses are a leguminous crop like dry grains, dry bean, pea, lentil, chickpea and fava-bean etc." The FAO (1994) incorporates 11 essential pulses and avoids oil-crop legume seeds (soybean and nut) and those collected green as fresh vegetables (green peas and green beans).

Pulses are described by their one of a kind healthy benefit as wellspring of proteins, carbohydrates, fiber, nutrients, minerals, and phytochemicals with recognized health advantages. Pulses are eaten cooked, broiled, sprouted, and fermented and establish a critical piece of the day by day diet of the vast majority of the total populace. Pulse derived fermented products change broadly in many regions of the world because of they are well constitute nutritional property.

Biochemical changes in nutritional composition of fermented product caused to the microorganism are involved the indigenous microbiota present on the leguminous crop surface or the additional starter cultures. Fermentation improves pulses flavour, surface, appearance, shelf-life, supplement digestibility, and nourishing quality. Moreover, this procedure diminishes non dietary compounds present in legume seeds, for example, protease inhibitors, oligosaccharides, phytate, and lectins.

Fermented legumes are at present getting a lot of consideration for their wellbeing advancing properties and malady forestalling impacts. Although the greater part of the human-controlled preliminaries directed up to this point have been centred around soybean products (Sugano, 2005), there are numerous other fermented pulse products wherein characteristics are gotten from their supplement and phytochemical constituents and the probiotic characteristic of microorganisms engaged with fermentation. In this unique circumstance, the goal of this chapter is to portray the wholesome composition of pulses, the biochemical changes happening during fermentation, and the emerging proof indicating their latent capacity impact on chronic disease avoidance.

Fermentation and fermented pulse Foods in developing nations

Pulses are processing in developing nations for the considerable long time for utilizing customary processing strategies of granulating, fermentation, soaking, germination and dehulling is utilization for further use. Other novel food processing including micronization, microwave handling, high pressure processing (HPP), pulse electric field (PEF), irradiation, and extrusion techniques have discovered potential use and application for pulse processing. fermentation remains to a great extent significant for pulse processing and enlarging elaborate review because based on its owned improved functionalities, increment dietary composition, and production of bioactive component (Frais J et at 2016).

Fermentation is a generally represent as a processing technique used to change over substrates into new products completed the activity of microorganisms (Adebiyi JA et at 2016). Fermentation is additionally utilized in a more extensive sense for the deliberate utilization of microorganisms to get useful products for people on a industrial scale. Such industrial products may incorporate biomass, catalysts, essential and secondary metabolites, recombinant, and biotransformation products. The biochemical changes that proceed all the food fermentation process prevail to the adjustment of the substrate (starch or sugar) and producing of different compounds, (for example, alcohol and acid). Fermentation, has been enhance the surface, appearance, colour, flavour, shelf life, and furthermore protein digestibility of pulses. It further diminishes the presence of "antinutritional factors" including phytate, lectins, oligosaccharides, and protease inhibitors (Adebiyi JA et at 2016, Frais J et at 2016). Particularly in rural and conventional networks, unconstrained fermentation is generally utilized for pulse processing. In any case, better and improved fermentation methods regarding explicit strain advancement have been urged and acquainted with improve product and nourishing quality, microbial wellbeing, and product yield. Notwithstanding fermentation, other processing tasks could include baking, cooking, and compositing, among others.

Microbiology of pulse fermentation

The microbiota of fermented pulses-based food is to a great extent subordinate with respect to temperature, pH, water activity, kind of substrate, and salt levels. The three significant sorts of microorganisms used at the time of fermentation of pulses are microscopic organisms of the Bacillus species, lactobacillus species; some fungal species; and perhaps yeasts. In prevailing element of these pulses based fermented foods, the fermentation procedure is unconstrained (common), and that's way a blend of microorganisms may act as the parallel or successively. This may subsequently cause changing and nonconsistent products and possible production of pathogenic microorganisms and toxins. Nevertheless, lactic acid microbes are predominant, typically fussy and develop energetically in most food substrates lessening the pH quickly to a point where other competing organisms are never again ready to develop (Satish KR et at 2013). A few industrial fermentations have additionally applied lactic acid microbes for the production of functional foods and compounds/metabolites. For a long time, indigenous or conventional fermented foods have formed a fundamental part of the eating system and might be set up in the cottage industry utilizing straightforward procedures and household gear (Aidoo KE et at 2006).

Fermented pulse-based foods are increasingly extensive and usable in developing nations, particularly in India where as it is proceed on as competitive modification in the networks of specific families, a training ensured by custom.

Fermentation of pulses and other food crops is related with decrease of pH; changes in carbs (starch, fibers, saccharides, sugars), proteins (amino acids), and lipids; "antinutritional" factors; and enzymatic debasement of various compounds. It also prompts the improvement of surface, taste, and smell of the final product. from different adjustments, fermentation of pulses is also connected with the arrangement of compounds because of microbial activities on endogenous compounds. Such compounds incorporate alcohols, ketones, natural acids, and aldehydes that further add to the smell associated with fermented pulse-based foods.

Advancement of Super Food from Fermented Pulses

Pulses based fermented Products	Name of microbes for using	Raw material	Nature of fermentation/sensory attribute	Regions
Kinema	Bacillus subtilis; B. licheniformis, B. circulans, B. thuringiensis and Bacillus sphaericus. Bacillus, Enterococcus faccium, Candida parasilosis	Yellow seeded soybean	Alkaline fermentation, sticky	Darjeeling or Sikkim.
Tungrymbai	Bacillus subtilis, LAB and yeast.	Soybeans.	Alkaline fermentation.	Mahalaya.
Axone	Bacillus subtilis.	Soybeans.	Alkaline, sticky batter	Northern east India.
Bekang	Bacillus subtilis.	Soybeans.	Alkaline fermentation.	Mizoram.
Masyaura	Pediococcus pentosaceous, Pediococcus acidilactic, and Lactobacillus sp. Saccharomyces cerevisiae and Candida versatilis, Cladosporium sp., Penicillium sp. and Aspergillus niger.	Blackgram or green gram,	Dry, ball-like, brittle, condiment	Darjeeling hills and Sikkim.
Wari	Lactobacillus bulgaricus and Streptococcus thermophilus.	Black bean and soybeans.	Ball-like, brittle, side dish	Uttar Pradesh.
Sufu	Actinomucor elenans, Mu. silvatixus, Mu. corticolus, Mu. hiemalis, Mu. praini, Mu. racemosus, Mu. subtilissimus, Rhiz. chinensis	Soybean curd	Mild-acidic, soft	China, Taiwan
Sieng	Bacillus sp	Soybeans	Alkaline, sticky	Cambodia, Laos
Peruyyan	Bacillus subtilis, Bacillus sp. and Lactic Acid Bacteria	Soybeans	Alkaline fermentation, sticky	Arunachal Pradesh.
Uri	Bacillus spp.	Locust bean	Alkaline, sticky, condiment, soup	Western Africa
Natto	B. subtilis (natto)	Soybeans	Alkaline, sticky	Japan

Nourishment innovation by fermentation of pulse-based food

Fermentation is an appropriate procedure to change over and improve the nature of a raw and poor digestible legumes into increasingly adequate, satisfactory, protected, nutritive, and healthy value included palatable products. The microorganisms associated with legume fermentation hydrolyze and utilize seed constituents transfer about the processing of derived-valuable products and produce antimicrobial compound and alluring natural acids can be preserve the food from the concealment of the development and endurance of undesirable microflora. Fermentation gives a few favourable circumstances over other conventionally possible techniques for legume processing, notwithstanding being more affordable.

Innovation in protein and amino acids during fermentation

Fermented food pulses significant wellspring of proteins, e.g., from 12–18% in dosa and idli products to 40% in tempeh (Krishnamoorthy et al., 2013; Starzyńska-Janiszewska et al., 2015). The positive changes in protein quality of pulses during lactic acid fermentation have been estimate by a few analysts. In the preparation of dosa, fermented black gram prompted a slight increment in the proteinase activity that achieves an expansion in the complete nitrogen, dissolvable proteins, and well-adjusted amino acid batter with 50% of basic amino acids over total amino acids (Balasubramanian et al., 2015). The replacement of black gram by mung bean produced more nutritious batters as far as complete nitrogen, protein, and free amino acids. Idli products give larger limiting amino acid aggregate and invitro protein digestibility than unfermented seeds (Riat and Sadana, 2009). Moreover, the fermented active yeast assortments of cowpeas, peas, and kidney beans improved the protein compound score and basic amino acid index, adding to the general protein quality. Lactic acid fermentation of faba bean prompted expanded amounts of essential amino acids and the hypotensive γ-aminobutyric acid (GABA), and enhanced the invitro protein digestibility. Also, during the processing of ugba with kidney beans, the substance of basic amino acids was improved to meet FAO dietary necessities (Audu and Aremu, 2011). Correspondingly, chickpeas, common beans, and bambara groundnut tempeh foods showed higher in vitro protein digestibility, total amino-acid content, accessible Lys, just as the determined protein productivity proportion than crude flour (Bujang and Taib, 2014; Reyes-Bastidas et al., 2010;). Also, certain strains of Rhizopus showed higher proteolytic activity by discharging a few times more amino acids than different strains.

Protein inhibitors are generally decreased after pulse fermentation, contributing to the improvement of the general protein quality. Trypsin inhibitory activity was for the most part disposed of during the preparation of tempeh-like products from cowpeas, ground beans, and chickpeas, and a corresponding increment on protein digestibility. It has additionally been accounted for that lactic-acid microscopic organisms prompted imperceptible activity of trypsin inhibitors (Shimelis and Rakshit, 2008;

Coda et al., 2015), while fermented assortments of cowpeas, peas, and kidney beans with Saccharomyces cerevisiae caused equal degradation of 37% and 49% (Khattab and Arntfield, 2009).

Innovation in starch, carbohydrates, and dietary fiber during Fermentation

Pulses are a defensive, fibrous, unpalatable hull that represents to among 0.09–0.3% of dried seeds. Seed hull contains a insoluble fiber and this part can be evacuated when dehulled grains (dhals) are the starting material of fermented derived products. Subsequently, the dehulling procedure prompted an expansion in fiber solvency, palatability, digestibility and by and large nutritive quality (Nalle et al., 2010). Soaking and cooking are typically previous treatments to fermentation. Drenching adds to the hydration of the seeds and causes the leaching effect in dissolvable carbohydrates and heat treatment further incites the starch gelatinization and improves starch digestibility.

Fermented pulses give a significant wellspring of carbohydrates (50–70%) (Abu-Salem and Abou-Arab, 2011). Starch degradation is a complex biochemical procedure that is regulated by endogenous pulse enzymes and those gave by fermentative microorganisms. Endogenous and microbial amylases assume a significant role during fermentation of pulse and expanded activity in the early stage was indicated that declined bit by bit with the fermentation improvement. During the hydrolysing of starch, amylose and amylopectin decline step by step through the course of fermentation and diminishing sugars are produced (Audu and Aremu, 2011). In this specific circumstance, LAB glucose digestion prompted a pH drop related with lactic acid production.

Lactic acid fermentation strikingly builds the starch absorbability of pulses and tempeh-like products. Nonetheless, blended culture fermentation with Rhizopus oligosporus and Aspergillus oryzae produce less invitro bioavailability of sugars (Abu-Salem and Abou-Arab, 2011).

Fermented pulse products are perceived as a decent wellspring of RS (8–15%) (Granito and Álvarez, 2006). For example, chickpea tempeh contains RS levels three-fold higher than crude flour (Angulo-Bejarano et al., 2008). Idli products have been distinguished as one of the significant RS providers in Indian populaces. Distinctive data has been included for about the production of total dietary fiber (TDF), insoluble (IDF), and soluble (SDF) divisions in fermented pulses. A diminishing in SDF has been accounted for in fermented bengal grams, cowpeas, green grams, and black beans, while IDF demonstrated a noteworthy increment, adding to the expansion of TDF (Granito and Álvarez, 2006). In fermented lentils, the substance of neutral dietary fiber (NDF), cellulose, and hemicellulose drained, while lignin content expanded twice.

There is an enormous accumulation of information on the decrease of raffinose family oligosaccharides during the lactic acid fermented pulses, making the final products progressively worthy by easing fart distress and intestinal cramps (Granito et al., 2003;

Madodé et al., 2013; Martínez-Villaluenga et al., 2008; Shimelis and Rakshit, 2008). Comparative outcomes have been accounted for cowpea natto, where Bacillus subtilis prompted the absolute expulsion of raffinose, stachyose, and verbascose of pea, and invitro and invivo studies showed impressive fermentability exhaustion (Madodé et al., 2013). Fungal fermentation is less compelling in α-galactoside evacuation (Starzyńska-Janiszewska et al., 2014), while yeast fermentation expanded essentially the substance of raffinose, stachyose, and verbascose of peas, cowpeas, and kidney beans.

Innovation in lipids during fermentation

Impact of the fermentation procedure on the lipid content and profile in pulses are moderately rare. Prinyawiwatkul et al. (1996) examined the impact of fungal maturation in lipid substance and unsaturated fatty acid composition of cowpea-like tempeh. Lipid content expanded from 2.2% in unfermented flour to 2.8% after 24 hrs of fermentation and the significant UFA were linoleic, palmitic, and linolenic acids. Niveditha et al. (2012) found that the lipid substance of Canavalia assortments fermented with Rhizopus oligosporus experienced a slight increment and UFA prevailed over the immersed ones. Reyes-Moreno et al. (2004) revealed a half lipid decline in chickpea tempeh in correlation with unfermented flour.

Innovation in phytic acid and mineral bioavailability during fermentation

Legumes are the limitation of phytic acid substance, considered as essentially inferior amount of mineral bioavailability. Numerous endeavours have been made to decrease the measure of phytate in legumes, and fermentation has been recognized as one of the most effective remedy, this is because of acidic conditions occurring during fermentation that increase the phytase activity prompting phytic-acid degradation and therefore a larger mineral accessibility is accomplished (Yadav et al., 2013). Phytic-acid Decadence in legumes throughout fermentation relies upon numerous processing parameter (time and temperature). (Khattab and Arntfield, 2009). according to this above notice that traditional Indian idli and dosa give a 69% decrease of phytic-acid substance and an expansion in Ca and Fe accessibility (Krishnamoorthy et al., 2013). The Zn bioaccesibility in those fermented products expanded 71% and 50%, separately, and to a more extent the Fe bioaccesibility 277% and 127%, respectively (Hemalatha et al., 2007). These foods are all around acknowledged for Indian kids and are suggested for malnourishment (Dahiya et al., 2014). And solid-state fermentation enumerates to the decrease in phytic-acid substance, an impact that has been associated with a higher invitro protein digestibility (Abu-Salem and Abou-Arab, 2011).

Innovation in vitamins during fermentation

A few microorganisms can synthetize water-soluble nutrients, making fermented pulses progressively nutritious. However, natural lactic-acid fermentation of various lentils, kidney beans, and cowpeas prompted larger thiamin and riboflavin content.

Similarly, Indian dosa batters give more thiamin, riboflavin, and cobalamin content than unfermented products, and the substitution of black gram by mung beans showed higher vitamin B group substance. In any case, the fermentation of bambara groundnut with Rhizopus oligosporus decreased thiamin content, while riboflavin, folic acid, niacin, and biotin expanded significantly (Fadahunsi, 2009). Also, the substance of nutrient E expanded during natural fermentation of cowpeas (Doblado et al., 2003), while diminished marginally in fermented pigeon peas and lupins.

Innovation in phenolic compounds and antioxidant activity during fermentation

As of late, there has been expanding enthusiasm for exploring phenolic compounds due to their antioxidant activity with defensive wellbeing consequences for oxidative stress prompted illnesses (Sharma et al., 2011). Pulses are an astounding wellspring of phenolic compounds that are mainly accumulated in the hull. The dehulling procedure that typically is performed preceding fermentation prompted a sharp decrease in tannin content. A huge increment in (+)- catechin and hydroxybenzoi-acid substance was found in naturally fermented lentils. Fermentation of cowpeas with Lactobacillus plantarum brought about a decrease of conjugated types of ferulic and p-cumaric and hydroxycinnamic subsidiaries, the synthesis of tyrosol, and an expansion in complimentary quercetin due to hydrolysing of quercetin glucosides. Changes in the phenolic composition of pulses during the fermentation are attributed to glycosidases and esterases of LAB discharging free aglycones and phenolic acids (Esteban-Torres et al., 2015; Ferreira et al., 2013; Jiménez et al., 2014; Limón et al., 2014), and less esterified pro-anthocyanidins and hydroxycinnamic acids adding to their antioxidant properties (Esteban-Torres et al., 2015). Also, microbial phenolic acid decarboxylases and reductases permit the amalgamation of phenolic metabolites with antioxidant activity (Landete et al., 2015).

The antioxidant property of fermented pulses inferred foods related with phenolic compounds have been broadly reported (Moktan et al., 2011; Torino et al., 2013; Torres et al., 2006). Dhokla and idli products exhibited higher free-radical scavenging, metal-quelating, and lipid peroxidation inhibitory activities than their unfermented batter (Moktan et al., 2011). Lactic-acid fermentation of pigeon peas, bambara groundnuts, and kidney beans realized an expansion in the free-dissolvable polyphenols, lessened the substance of bound polyphenols, and therefore upgraded antioxidant activity. Bean and lentil tempeh exhibited higher soluble phenolic concentration, radical scavenging, and antioxidant activity than unfermented grains (Reyes-Bastidas et al., 2010; Torino et al., 2013). Based on this conclusion, the adjustments in compounds and synthesis of phenolic compounds in pulses propose that fermentation can be considered as a practicable vital procedure to promote health benefits and counteract oxidative stress.

Innovation in other minor bioactive compounds during fermentation

Fermentation of various pulses with LAB brings about a perceptible increment in GABA content, a free nonprotein amino acid present in low amounts in grains, while fermentation with Bacillus subtilis was fairly inferior (Limon et al., 2014; Torino et al., 2013). These outcomes propose that glutamic acid decarboxylase, the enzyme liable for the GABA biosynthesis, is more dynamic in LAB than in the Bacillus strains engaged with the fermentation procedure. elimination of saponins after natural and controlled lactic-acid fermentation of common bean and 58% decrease in saponins of soybean tempeh fermented with Rhizopus oligosporus. Fermented Lactic-acid has been appeared to diminish vicine and convi-cine, cyanogenic glycosides in faba beans (Coda et al., 2015), and the natural toxin β-ODAP from grass peas (Starzyńska-Janiszewska and Stodolak, 2011). Also, the capacity of LAB to degrade biogenic amines phenylethylamine, spermine, and spermidine, the fact that the histamine and tyramine were found in fermented lupin products at levels lower than those causing adverse wellbeing impacts (Bartkiene et al., 2015). Other minor compounds have additionally been depicted for pulses, however fermented seeds are scarce, and further studies has been expected to better comprehend their transformations during fermentation.

Health advantage of pulse fased fermented foods

Fermented pulse products and weigh management

Overweight and obese people are in danger for several medical conditions that contribute to morbidity and mortality, including diabetes, cardiovascular diseases, and other metabolic complications. An everyday energy unevenness prompts weight increase after some time and anticipation of overabundance weight addition can be accomplished with low vitality thickness food. Supplanting vitality thick foods with pulses can improve satiety (Rebello et al., 2014).

Fermented pulses gives a good source of nutritional property like proteins, peptides, and amino acids, which make them candidates to advance weight reduction by sensation of fullness (Iwashita et al., 2006; McKnight et al., 2010). the fact that studies on fermented pulse-derived products are as yet forthcoming, their composition suggests that they can adjust natural procedures that balance obesity.

Fermented pulse-products and diabetes

Pulses are elective food in dietary methodologies to oversee blood glucose levels. pulses give high measures of RS, for the most part characterized as starch products not digested in the small digestive system, which eventually leads to bring down GI and lower insulin obstruction (Messina, 2014). Lower GI and insulin obstruction are contributing components to decreasing both the rate and seriousness of type 2 diabetes, another variable engaged with metabolic disorder that is likewise connected with waistline fat testimony, dyslipidaemia, and hyperglycemia. Among traditional pulses, mung beans have been prescribed as a potential antidiabetic food for diabetic patients, and matured nourishments likewise help to diminish the predominance of diabetes in Asian populaces. Fermented mung bean products have been prescribed to oversee diabetes because of their low-GI, high-fiber substance, and phenolic compounds, which improve oxidative stress-induced hyperglycemia (Atkinson et al., 2008; Landete et al., 2015; Maiti and Majumdar, 2012; Randhir and Shetty, 2007). The antihyperglycemic impact of fermented mung bean extracts observed in an alloxan-incited diabetic mice model was ascribed to its higher GABA substance and free amino acids (Yeap et al., 2012). Fermented tropical legume-based regimens additionally exhibited a modulatory impact of the oxidative stress and protection is hepatic tissue harm in streptozotocin-incited diabetic rats, an impact credited to higher intake in phenolic antioxidants reinforcements (Ademiluyi and Oboh, 2012). While these outcomes show fundamental facts on the impact of fermented pulses on diabetes human clinical studies are urged to approve the outcomes observed in preclinical studies.

Fermented pulse-derived products and cardiovascular diseases

Normally fermented cowpeas exhibited antioxidant, cancer prevention agent and lipid-bringing properties that may contribute down to bringing down the danger of the improvement of cardiovascular ailments. Critical enhancements in plasma antioxidant and hepatic action of antioxidant reinforcement compounds were observed in albino Wistar rats nourished with cowpea- fermented flours for 14 days. Moreover, liver weight and plasma cholesterol and triglyceride levels were positively influenced (Kapravelou et al., 2015). These impacts were ascribed to dietary fiber components (Anderson and Major, 2002; Bazzano, 2008; Ma et al., 2008) as well as flavonoid uptake (Cassidy et al., 2011). Other minor segments, for example, saponins, oligosaccharides, and phytosterols can also add to hinder the intestinal retention of cholesterol.

Some fermented-legume products are bringing down the risk of cardiovascular disorder due to their circulation of blood pressure bringing down impacts. One recently report says that, navy bean milk fermented by Lactobacillus bulgaricus and Lactobacillus plantarum B1-6 displayed angiotensin-I changing over protein (ACE) inhibitory action, an impact that was ascribed to the presence of bioactive peptides (Rui et al., 2015). Thus, the fermentation of mung bean milk by L. plantarum B1-6 brought about the arrival of smaller and progressively hydrophilic peptides with higher ACE inhibitory action (Wu et al., 2015). Fermented adzuki beans with Lactococcus lactis and Lactobacillus rhamnosus and lactic-acid fermented lentils and kidney beans either naturally or by L. plantarum CECT 748 (Torino et al., 2013) exhibited observable aggregation of GABA, which may add to the antihypertensive impact related with these fermented foods (Franciosi et al., 2015; Suwanmanon and Hsieh, 2014).

Antioxidant agents additionally add to the cardioprotective impact of fermented legumes as it has been appeared during the oral organization of 50% ethanol concentrates of red bean natto to Sprague–Dawley rodents (Chou et al., 2008). The presence of natto kinase in natto adds another cardio preventive credit to these products since this extracellular catalyst has fibrinolytic action and accordingly have been considered effective against thrombolytic scenes. Other important focal points depicted for natto kinase incorporate treatment of hypertension, Alzheimer's disease, and vitreoretinal issue (Dabbagh et al., 2014).

Fermented pulse-derived products and cancer

At present, the Food and Drug Administration (FDA), Canadian Cancer Society (CCS), and the World Cancer Research Fund (WCRF) specify the uses of pulses and to lessen the malignant cancer risk. These are also add to the specify regular intake of 25 g of non-starch polysaccharide to help arrive at public wellbeing goals. (WCRF/AICR, 2010). Fermented food utilization may present an assortment of significant dietary and remedial advantages to improve the advancement of malignant cancer (Kandasamy et al., 2011). The potential anticancer impact of tempeh-like concentrates from Canavalia cathatica and C. maritima on colon malignant cancer cell lines MCF-7 and HT-29 has been illustrated (Niveditha et al., 2013). Fermented mung bean extracts have indicated chemo preventive activity on 4T1 breast cancer cells through the incitement of insusceptibility, lipid peroxidation, and modulation of aggravation (Yeap et al., 2013).

Minor segments of heartbeats, for example, soyasaponins are hydrolysed by LAB discharging aglycones, for example, soyasapogenol An and B wherein the lipophilic center have been recognized as being answerable for cell passing of colon-malignant growth cells (Gurfinkel and Rao, 2003). Bowman–Birk protease inhibitors have displayed anticancer action through hindrance of protease-intervened aggravation and development of human colon-disease cells (Clemente and Arques, 2014). Vegetable lectins are likewise considered as antitumor specialists.

Fermented pulse-derived products in healthy aging and stress

Notwithstanding disease anticipation, pulses consumption has been related with longevity and with potential improvement of psychological wellness in maturing. An investigation of five companions of older individuals (≥70 years) recognized high legumes intake as a consistent and critical defensive dietary component among nine significant gatherings of food (Darmadi-Blackberry et al., 2004). pulses utilization is related with decreased pressure, tension emotional, distress, and somatic symptoms in older adults (Smith, 2012). Fermentation product of mung bean is fermented by the different Rhizopus species (fungi) exhibited by powerful anti-inflammatory and antinociceptive activities in a dose-dependent manner in invitro (Ali et al., 2014). notwithstanding the above basic systems these impacts is not completely comprehended, the author ascribed that the accumulate of GABA, total essential amino-acids, and phenolic antioxidants. Moreover, fermented mung beans products have recently been documented to contribute to the alleviation of heatstroke occurring under stress conditions in vivo (Yeap et al., 2014). All concept are examine that the , these are a starter think about and further research ought to be directed to promote and improve the processing of fermented pulses in healthy aging and prosperity.

Fermented pulse-derived products as probiotic vehicle

LAB assume an essential part in the formulation of fermented pulses foods. Their mix with a lot of non-starch polysaccharides incite symbiotic potential advantages by the arrangement of different acidic compounds, for example, acetic acid, lactate, butyrate, and propionate, and the arrival of short-chain unsaturated fats with further decline of pH related with positive adjustment in the gastrointestinal microecology (Parvez et al., 2006). Fermented mung bean milk prepared with Lactobacillus plantarum was proposed as a probiotic carrier with health advantages. In like manner, Rhizopus fermentation prompts the abatement of pH, and it has been recommended that tempeh-like products can upgrade the steadiness of intestinal valuable microorganisms (Dinesh Banu et al., 2009). The advantages of lower pH in the digestive tract incorporate upgraded multiplication and endurance of beneficial microorganisms while together inhibit the development of unwanted pathogens. In any case, further examinations ought to be urged to determine the act of fermented-pulse foods on gastrointestinal health.

Future possibilities of pulses-based food

Attributable to their relative accessibility, pulses are perceived as noteworthy sources of food.

All things considered; they are viewed as "food for poor people" in most creating countries. Fermentation as a food processing technique can improve the quality and other wellbeing advancing advantages of pulses. utilization of fermented pulses-based food would therefore be advantageous and generally add to sustenance and safety of food. Whereas, these fermented pulses-based foods are promptly accessible, and every day per capita utilization in customary settings has been declining in the last few years, and these are connected with interdependent in expansion of chronic diseases plaguing both developing and developed nations. While a portion of the innate bioactive mixes in fermented pulses-based food might repress nutrient accessibility, fermentation can successfully diminish their capacity to do this, in this manner guaranteeing that the bioactive mixes present give some functional activities.

There is a current potential market for functional foods, but still now approachability of shelf-stable items can obstruct their possibilities. In that capacity, mechanisms to ensure access to innovation and aptitude among nearby and small-scale food processors ought to be upgraded. while cost may retard the arrangement of industrially susceptible starter culture, conveyance of such starter culture for improved and compelling fermentation could be accomplished utilizing of dehydrated fermented foods (with suitable fermenting organisms), for consequent use. In particular expanding familiarity with pulses and subsequent fermented products from such crops as a source of functional and health-promoting foods determine the act of government, nongovernmental associations, and alternative applicable stakeholders within the well-being and other related segments. This will to a huge extent ensure that developing nations accomplish the truly necessary and visualized food and nutrition security.

Conclusion

Most fermented nourishments are regarded as sheltered to eat. Lactic acid microbes produce lactate and acetic acid, which lead to down pH of the nourishment and restrain other pathogenic organism. Ethanol, hydrogen peroxide, and protein-based toxin given off by bacteria to prevent the growth of bacteria like antibacterial effect. *can also* be created and help in safeguarding and security. Numerous customary Asian aged nourishments depend more on development and compound creation by *Aspergillus* spp. of which strains are realized which can create strong mycotoxins. However, numerous strains utilized in fermented nourishment generation seem to have been 'trained' more than a large number of years. Later ordered examinations recommend that *Aspergillus oryzae* which is utilized in a few fermented nourishments is a non-toxigenic trained assortment of A. flavus Fermentation of nourishment has developed from a characteristic natural art through the using chose and controlled conditions and starter cultures to logical seeing, yet a few parts of this biotechnological advancement are not yet completely comprehended, including the choice of mycotoxin free strains of molds for matured nourishment production. The following stage are the development and utilization of gene technology to improve certain ideal attributes. Like, that have recently been endorsed for use in the UK incorporate a hereditarily adjusted, improved strain of baker's yeast, Saccharomyces cerevisiae to improve carbon dioxide gas during the production in bread processing dough and the utilization of a bacterially-inferred chymosin to supplant calves' rennet in cheddar production.

References

A.M.O. Leite, M.A. Miguel, R.S. Peixoto, A.S. Rosado, J.T. Silva, V.M. (2013). PaschoalinMicrobiological, technological and therapeutic properties of kefir: A natural probiotic beverage *Braz. J. Microbiol.*, 44, pp. 341-349

Abu-Salem, F.M., Abou-Arab, E.A., (2011). Physicochemical properties of tempeh produced from chickpea seeds. *Journal of American Science* 7, 107–118.

Adebiyi JA, Obadina AO, Adebo OA, Kayitesi E. (2016). Fermented and malted millet products in Africa: Expedition from traditional/ethnic foods to industrial value added products. Critical Reviews in Food Science and Nutrition. 2016. DOI: 10.1080/ 10408398.2016.1188056

Adebo OA, Njobeh PB, Adebiyi JA, Gbashi S, Phoku JZ, Kayitesi E (2017a) Fermented pulse-based foods in developing nations as sources of functional foods. In: Hueda MC (ed) Functional food-improve health through adequate food.

Ademiluyi, A.O., Oboh, G., (2012). Attenuation of oxidative stress and hepatic damage by some fermented tropical legume condiment diets in streptozotocin-induced diabetes in rats. *Asian Pacific Journal of Tropical Medicine* 5, 692–697.

Ahmed, A. R. and Ramanathan, G., (1988) Effect of natural fermentation on the functional properties of protein enriched composite flour, *J. Food Sci.*, 53, 218.

Ahren IL, Xu J, Onning G, Olsson C, Ahrne S, Molin G (2014) Antihypertensive activity of blueberries fermented by Lactobacillus plantarum DSM 15313 and effects on the gut microbiota in healthy rats. *Clin Nutr* 34:719–726

Aidoo KE, Nout NJR, Sarkar PK. (2006) Occurrence and function of yeasts in Asian indigenous fermented foods. FEMS Yeast Research. ;1:30-39. DOI: 10.1111/j.1567-1364.2005.00015.x

Ali, M. A. M., El-Tinay, A. H., & Abdalla, A. H. (2003). Effect of fermentation on the in vitro protein digestibility of pearl millet. Food Chemistry, 80, 51–54. https://doi.org/10.1016/S0308-8146(02)00234-0

Ali, N.M., Mohd Yusof, H., Yeap, S.K., Ho, W.Y., Beh, B.K., Long, K., Koh, S.P., Abdullah, M.P., Alitheen, N.B., (2014). Antiinflammatory and antinociceptive activities of untreated, germinated, and fermented mung bean aqueous extract. Evidence-Based Complementary and Alternative Medicine.

Alka, S., Neelam, Y., & Shruti, S. (2012). Effect of fermentation on physicochemical properties & in vitro starch and protein digestibility of selected cereals. *International Journal of Agricultural and Food Science*, 2, 66–70.

Allison, G. E., & Klaenhammer, T. R. (1998). Phage resistance mechanisms in lactic acid bacteria. *International Dairy Journal*, 8, 207–226.

Anderson, J.W., Major, A.W., 2002. Pulses and lipaemia, short- and long-term effect, potential in the prevention of cardiovascular disease. *British Journal of Nutrition* 88, S263–S271.

Atkinson, F.S., Foster-Powell, K., Brand-Miller, J.C., (2008). International tables of glycemic index and glycemic load values: 2008. *Diabetes Care* 31, 2281–2283.

Audu, S.S., Aremu, M.O., 2011. Effect of processing on chemical composition of red kidney bean (*Phaseolus vulgaris* L.) flour. *Pakistan Journal of Nutrition* 10, 1069–1075.

Balasubramanian, S., Jincy, M.G., Ramanathan, M., Chandra, P., Deshpande, S.D., (2015). Stud¬ies on millet idli batter and its quality evaluation. *International Food Research Journal* 22, 139–142.

Bartkiene, E., Krungleviciute, V., Juodeikiene, G., Vidmantiene, D., Maknickiene, Z., (2015). Solid state fermentation with lactic acid bacteria to improve the nutritional quality of lupin and soya bean. *Journal of the Science of Food and Agriculture* 95, 1336–1342.

Bazzano, L.A., (2008). Effects of soluble dietary fiber on low-density lipoprotein cholesterol and coronary heart disease risk. *Current Atherosclerosis Reports* 10, 473–477.

Blandino, A. - Al-Aseeri, M. E. - Pandiella, S. S. Cantero, D. - Webb, C.: Cereal-based fermented foods and beverages. *Food Research International*,36, 2003, pp. 527-543.

Bujang, A., Taib, N.A., (2014). Changes on amino acids content in soybean, garbanzo bean and groundnut during pre-treatments and tempe making. *Sains Malaysiana* 43, 551–557.

Campbell-Platt, G. (1987) Fermented Foods of the World: a Dicrionary and Guide. Butterworths, London

Caplice E, Fitzgerald GF (1999) Food fermentations: role of microorganisms in food production and preservation. *Int J Food Microbiol* 50:131–149

Cassidy, A., O'Reilly, É.J., Kay, C., Sampson, L., Franz, M., Forman, J.P., Curhan, G., Rimm, E.B., 2011. Habitual intake of flavonoid subclasses and incident hypertension in adults. *American Journal of Clinical Nutrition* 93, 338–347.

Chandrasekhar, A. and Desikachar, H. S. R., (1983). Sorghum quality studies. I. Rolling quality of sorghum dough in relation to some physiochemical properties, *J. Food Sci. Technol.*, 20, 281.

Chaves-Lopez, C., Serio, A., Grande-Tovar, C. D., Cuervo-Mulet, R., Delgado-Ospina, J., & Paparella, A. (2014). Traditional fermented foods and beverages from a microbiological and nutritional perspective: The Colombian Heritage. Comprehensive Reviews in Food Science and Food Safety, 13, 1031–1048. https://doi. org/10.1111/1541-4337.12098

Chou, S.T., Chao, W.W., Chung, Y.C., (2008). Effect of fermentation on the antioxidant activity of red beans (*Phaseolus radiatus* L. var. Aurea) ethanolic extract. *International Journal of Food Science and Technology* 43, 1371–1378.

Clemente, A., Arques, M.C., (2014). Bowman-Birk inhibitors from legumes as colorectal chemopre¬ventive agents. *World Journal of Gastroenterology* 20, 10305–10315.

Coda, R., Melama, L., Rizzello, C.G., Curiel, J.A., Sibakov, J., Holopainen, U., Pulkkinen, M., Sozer, N., (2015). Effect of air classification and fermentation by Lactobacillus plantarum VTT E-133328 on faba bean (*Vicia faba* L.) flour nutritional properties. *International Journal of Food Microbiology* 193, 34–42.

Dabbagh, F., Negahdaripour, M., Berenjian, A., Behfar, A., Mohammadi, F., Zamani, M., Irajie, C., Ghasemi, Y., 2014. Nattokinase, production and application. Applied Microbiology and Bio¬technology 98, 9199–9206.

Dahiya, P.K., Nout, M.J.R., van Boekel, M.A., Khetarpaul, N., Grewal, R.B., Linnemann, A., (2014). Nutritional characteristics of mung bean foods. *British Food Journal* 116, 1031–1046.

Darmadi-Blackberry, I., Wahlqvist, M.L., Kouris-Blazos, A., Steen, B., Lukito, W., Horie, Y., Horie, K., (2004). Legumes, the most important dietary predictor of survival in older people of different ethnicities. *Asia Pacific Journal of Clinical Nutrition* 13, 217–220.

Day, C. N., & Morawicki, R. O. (2018). Effects of fermentation by yeast and amylolytic lactic acid bacteria on grain sorghum protein content and digestibility. *Hindawi Journal of Food Quality*, 2018, 1–8. https://doi.org/10.1155/2018/3964392

Dinesh Banu, P., Bhakyaraj, R., Vidhyalakshmi, R., (2009). A low cost nutritional food "tempeh"- a review. *World Journal of Dairy & Food Sciences* 4, 22–27.

Dirar, M. (1993) The Indigenous Fermented Foods of rhe Sudan. CAB International. University Press, Cambridge

Divyashree, S., Sandeep, P. G., Bhavya, S. N., & Prakash, J. (2013). Effect of fermentation of wheat dough on nutritional quality of chapathis. In Proceedings of International Conference on Technological Advances in Superfoods for Health Care. International Institute of Food and Nutritional Sciences, Pondicherry, India. May 3–4.

Dlamini, N. R., Taylor, J. R. N., & Rooney, L. W. (2007). The effect of sorghum type and processing on the antioxidant properties of African sorghum based foods. Food Chemistry, 105(4), 1412–1419.

Doblado, R., Frias, J., Muñoz, R., Vidal-Valverde, C., (2003). Fermentation of Vigna sinensis var. carilla flours by natural microflora and Lactobacillus species. *Journal of Food Protection* 66, 2313–2320.

Dordevic, T., Marinkovic, S. S., & Dimitrijevic-Brankovic, S. I. (2010). Effect of fermentation on antioxidant properties of some cereals and pseudo cereals. *Food Chemistry*, 119, 957–963. https://doi.org/10.1016/j.foodchem.2009.07.049

Doudu, K. G., Taylor, J. R. N., Belton, P. S., & Hamaker, B. R. (2003). Factors affecting sorghum protein digestibility. *Journal of Cereal Science*, 38, 117–131. https://doi.org/10.1016/S0733-5210(03)00016-X

Dykes, L., & Rooney, L. W. (2006). Sorghum and millet phenols and antioxidants. *Journal of Cereal Science*, 44, 236–251.

Egli, I., Davidsson, L., Juillerat, M. A., Barclay, D., & Hurrell, R. F. (2002). The influence of soaking and germination on the phytase activity and phytic acid content of grains and seeds potentially useful for complementary feeding. *Journal of Food Science*, 67, 3484–3488. https://doi. org/10.1111/j.1365-2621.2002.tb09609.x

Egli, I., Davidsson, L., Juillerat, M. A., Barclay, D., & Hurrell, R. F. (2003). Phytic acid degradation in complementary foods using phytase naturally occurring in whole grain cereals. *Journal of Food Science*, 68, 1855–1859. https://doi.org/10.1111/j.1365-2621.2003

El-Hag, M. E., El-Tinay, A. H., & Yousif, N. E. (2002). Effect of fermentation and dehulling on starch, total polyphenols, phytic acid content and in vitro protein digestibility of pearl millet. *Food Chemistry,* 77, 193–196. https://doi.org/10.1016/S0308-8146(01)00336 3

Emambux, M. N., & Taylor, J. N. (2003). Sorghum kafirin interaction with various phenolic compounds. *Journal of the Science of Food and Agriculture,* 83, 402–407. https://doi. org/10.1002/(ISSN)1097-0010

Esteban-Torres, M., Landete, J.M., Reverón, I., Santamaría, L., de las Rivas, B., Muñoz, R., (2015). A Lactobacillus plantarum esterase active on a broad range of phenolic esters. *Applied and Environmental Microbiology* 81, 3235–3242.

Fadahunsi, I.F., (2009). The effect of soaking, boiling and fermentation with Rhizopus oligosporus on the water soluble vitamin content of bambara groundnut. *Pakistan Journal of Nutrition* 8, 835–840.

FAO, (1994). Definition and Classifications of Commodities, Pulses and Derived Products. Online. Available from: http://www.fao.org/es/faodef/fdef04ee.htm.

Ferreira, L.R., Macedo, J.A., Ribeiro, M.L., Macedo, G.A., (2013). Improving the chemopreven¬tive potential of orange juice by enzymatic biotransformation. *Food Research International* 51, 526–535.

Fleming, S. E., Sosulski, F. W., Kilara, A., and Humbert, E. S., (1974) Viscocity and water absorption characteristics of slurries of sunflower and soybean flours, concentrates and isolates, *J. Food Sci.*, 39, 188.

Frais J, Penas E, Martinez-Villaluenga C. (2016)Fermented pulses in nutrition and health promotion. In: Frias J, Martinez-Villaluenga C, Penas E, editors. Fermented Foods in Health and Disease Prevention. Academic Press; Rome, Italy: Food and Agriculture Organization of the United Nations. pp. 385-416. DOI: 10.1016/B978-0-12-802309-9.00016-9

Franciosi, E., Carafa, I., Nardin, T., Schiavon, S., Poznanski, E., Cavazza, A., Larcher, R., Tuohy, K.M., (2015). Biodiversity and γ-aminobutyric acid production by lactic acid bacteria isolated from traditional alpine raw cow's milk cheeses. BioMed Research International 2015.

Golzarand, M., Mirmiran, P., Bahadoran, Z., Alamdari, S., & Azizi, F. (2014). Dietary phytochemical index and subsequent changes of lipid profile: A 3-year follow-up in Tehran Lipid and Glucose Study in Iran. ARYA Atherosclerosis, 10, 203–210.

Granito, M., Álvarez, G., (2006). Lactic acid fermentation of black beans (Phaseolus vulgaris), microbiological and chemical characterization. *Journal of the Science of Food and Agriculture* 86, 1164–1171.

Granito, M., Champ, M., Guerra, M., Frias, J., (2003). Effect of natural and controlled fermentation on flatus-producing compounds of beans (*Phaseolus vulgaris*). *Journal of the Science of Food and Agriculture* 83, 1004–1009.

Guermani, L., Villaume, C., Bau, H. W., Chandrasiri, V., Nicolas, J. P., & Mejean, L. (1992). Composition and nutritional value of okara fermented by Rhizopus oligosporus. Sciences des Aliments, 12, 441–451.

Gunness, P., & Gidley, M. J. (2010). Mechanisms underlying the cholesterol lowering properties of soluble dietary fiber polysaccharides. *Functional Foods,* 1, 149–155. https://doi. org/10.1039/c0fo00080a

Gupta, U., Rudramma, Rati, E. R., & Joseph, R. (1998). Nutritional quality of lactic fermented bitter gourd and fenugreek leaves. *International Journal of Food Science and Nutrition*, 49, 101–108.

Gurfinkel, D.M., Rao, A.V., (2003). Soyasaponins: the relationship between chemical structure and colon anticarcinogenic activity. *Nutrition and Cancer* 47, 24–33.

Han X, Yang Z, Jing X, Yu P, Zhang Y, Yi H, Zhang L (2016) Improvement of the texture of yogurt by use of exopolysaccharide producing lactic acid bacteria. BioMed Res Int. https://doi.org/10.1155/2016/7945675

Harland, B. F., & Harland, J. (1980). Fermentative reduction of phytate in rye, white, and whole wheat breads. *Cereal Chemistry*, 57, 226–229.

Hemalatha, S., Platel, K., & Srinivasan, K. (2007). Influence of germination and fermentation on bioaccessibility of zinc and iron from food grains. *European Journal of Clinical Nutrition*, 61, 342–348. https://doi.org/10.1038/sj.ejcn.1602524

Hepner G, Fried RR, Jeor SS, Fuseti L & Morin R (1979). Hypercholesteremic effect of youghur and milk. *American Journal of Clinical Nutrition*, 32, 19

Holck A, Heir E, Johannessen T, Axelsson L (2015) North European products. In: Toldra F (ed) Handbook of fermented meat and poultry, 2nd edn. Wiley Blackwell, West Sussex, pp 313–320

Hotz, C., & Gibson, R. S. (2007). Traditional food-processing and preparation practices to enhance the bioavailability of micronutrients in plants-based diets. Journal of Nutrition, 137, 1097–1100. https://doi. org/10.1093/jn/137.4.1097

Hubert, J., Berger, M., Nepveu, F., Paul, F., & Dayde, J. (2008). Effects of fermentation on the phytochemical composition and antioxidant properties of soy germ. Food Chemistry, 109, 709–721. https://doi.org/10.1016/j.foodchem.2007.12.081

Hugenholtz, J. - Smid E. D.: Nutraceutical production with food grade microorganisms. Current Opinions in Biotechnology, 13, 2002, pp. 497-507.

Iwashita, S., Mikus, C., Baier, S., Flakoll, P.J., (2006). Glutamine supplementation increases post¬prandial energy expenditure and fat oxidation in humans. *Journal of Parenteral and Enteral Nutrition* 30, 76–80.

Javanainen, P. - Linko, Y. Y. (1993). Factors affecting rye sour dough fermentation with mixed culture pre-ferment of lactic and propionic acid bacteria. *Journal of Cereal Science*, 18, 1993, pp. 171-185.

Jespersen, L.: Occurrence and taxonomical characteristics of strains of Saccharomyces cerevisiae predominant in African indigenous fermented foods and beverages. FEMS Yeasts Research, 3, 2003, pp. 191-200.

Jiménez, N., Esteban-Torres, M., Mancheño, J.M., De las Rivas, B., Muñoz, R., 2014. Tannin deg¬radation by a novel tannase enzyme present in some Lactobacillus plantarum strains. *Applied and Environmental Microbiology* 80, 2991–2997.

Kamiya S, Owasawara M, Arakawa M, Hagimori M (2013) The effect of lactic acid bacteria fermented soybean milk products on carrageenan-induced tail thrombosis in rats. *Biosci Microbiota Food Health* 32:101–105

Kandasamy, M., Bay, B.H., Lee, Y.K., Mahendran, R., (2011). Lactobacilli secreting a tumor anti¬gen and IL15 activates neutrophils and dendritic cells and generates cytotoxic T lymphocytes against cancer cells. *Cellular Immunology* 271, 89–96.

Kapravelou, G., Martínez, R., Andrade, A.M., López Chaves, C., López-Jurado, M., Aranda, P., Arrebola, F., Cañizares, F.J., Galisteo, M., Porres, J.M., (2015). Improvement of the anti¬oxidant and hypolipidaemic effects of cowpea flours (*Vigna unguiculata*) by fermentation, results of in vitro and in vivo experiments. *Journal of the Science of Food and Agriculture* 95, 1207–1216.

Kaur, I. P. - Chopra, K. - Saini, A. (2002) Probiotics: potential pharmaceutical applications. *European Journal of Pharmaceutical Science*, 15, 1-9.

Khan, M. N., Rhee, K. C., Rooney, L. W., and Cater, C. M., (1975). Bread baking properties of aqueous processed protein concentrates, *J. Food Sci.*, 40, 580.

Khattab, R.Y., Arntfield, S.D., Nyachoti, C.M., (2009). Nutritional quality of legume seeds as affected by some physical treatments: part 1, protein quality evaluation. *LWT - Food Science and Technology* 42, 1107–1112.

Kilara, A., Humbert, E. S., and Sosulski, F. W., Nitrogen extractability and moisture absorption characteristics of sunflower seed products, J. Food Sci., 37, 771, 1972.

Kim, H. Y. Min, J. H. Lee, J. H. Ji, G. E. (2000). Growth of lactic acid bacteria and bifidobacteria invnatural media using vegetables, seaweeds, grains andvpotatoes. Food Science and Biotechnology, 9, 2000,vpp. 322-324.

Kneifel, W., & Mayer, H. K. (1991). Vitamin profiles of kefirs made from milks of different species. *International Journal of Food Science & Technology*, 26, 423–428.

Krishnamoorthy, S., Kunjithapatham, S., Manickam, L., 2013. Traditional Indian breakfast (Idli and Dosa) with enhanced nutritional content using millets. *Nutrition and Dietetics* 70, 241–246.

Kruger, J., Taylor, J., John, R. N., & André, O. (2012). Effects of reducing phytate content in sorghum through genetic modification and fermentation on in vitro iron availability in whole grain porridges. *Food Chemistry*, 131, 220–224. https://doi.org/10.1016/j.foodchem.2011.08.063

Kuboye, A.O. (1985) Traditional fermented foods and beverages of Nigeria. In Development of Indigenous Fermented Foods and Food Technology in Africa. Proc. IFS/UNU Workshop, Douala, Cameroon, Oct. 1985. International Foundation for Science, Stockholm, Sweden

Landete, J.M., Hernández, T., Robredo, S., Dueñas, M., De Las Rivas, B., Estrella, I., Muñoz, R., (2015). Effect of soaking and fermentation on content of phenolic compounds of soybean (Glycine max cv. Merit) and mung beans (Vigna radiata [L] Wilczek). *International Journal of Food Sciences and Nutrition* 66, 203–209.

Landete, J.M., Hernández, T., Robredo, S., Dueñas, M., De Las Rivas, B., Estrella, I., Muñoz, R., (2015). Effect of soaking and fermentation on content of phenolic compounds of soybean (Glycine max cv. Merit) and mung beans (Vigna radiata [L] Wilczek). *International Journal of Food Sciences and Nutrition* 66, 203–209.

Leenhardt, F., Levrat-Verny, M., Chanlia, E., & Eameasy, C. (2005). Moderate decrease of pH by sourdough fermentation is sufficient to reduce phytate content of whole wheat flour through endogenous phytase activity. *Journal of Agricultural and Food Chemistry*, 53,98–102. https://doi.org/10.1021/jf049193q

Lesinski, G. B., Reville, P. K., Mace, T. A., Young, G. S., Ahn-Jarvis, J.,Thomas-Ahner, J., Clinton, S. K. (2015). Consumption of soy isoflavone enriched bread in men with prostate cancer is associated with reduced proinflammatory cytokines and immunosuppressive cells. Cancer Prevention Research, 8, 1036–1044. https://doi.org/10.1158/1940-6207.capr-14-0464

Liang, J., Han, B. Z., Nout, M. J. R., & Hamer, R. J. (2008). Effects of soaking, germination and fermentation on phytic acid, total and in vitro soluble zinc in brown rice. *Food Chemistry*, 110, 821–828. https://doi. org/10.1016/j.foodchem.2008.02.064

Limón, R.I., Peñas, E., Torino, M.I., Martínez-Villaluenga, C., Dueñas, M., Frias, J., (2014). Fermen¬tation enhances the content of bioactive compounds in kidney bean extracts. *Food Chemistry* 172, 343–352.

Lopetcharat K, Choi YJ, Park JW, Daeschel MA (2001) Fish sauce products and manufacturing: a review. *Food Rev Int* 17(1):65–88

Lorenzen, J. K., Nielsen, S., Holst, J. J., Tetens, I., Rehfeld, J. F., & Astrup, A. (2007). Effect of dairy calcium or supplementary calcium intake on postprandial fat metabolism, appetite, and subsequent energy intake. *American Journal of Clinical Nutrition*, 85, 678–687. https://doi.org/10.1093/ajcn/85.3.678

Ma, Y., Hébert, J.R., Li, W., Bertone-Johnson, E.R., Olendzki, B., Pagoto, S.L., Tinker, L., Rosal, M.C., Ockene, I.S., Ockene, J.K., Griffith, J.A., Liu, S., (2008). Association between dietary fiber and markers of systemic inflammation in the Women's Health Initiative Observational Study. Nutrition 24, 941–949.

Madodé, Y.E., Nout, M.J.R., Bakker, E.J., Linnemann, A.R., Hounhouigan, D.J., van Boekel, M.A.J.S., (2013). Enhancing the digestibility of cowpea (Vigna unguiculata) by traditional pro¬cessing and fermentation. *LWT - Food Science and Technology* 54, 186–193

Maiti, D., Majumdar, M., (2012). Impact of bioprocessing on phenolic content & antioxidant activ¬ity of mung seeds to improve hypoglycemic functionality. *International Journal of PharmTech Research* 4, 924–931.

Manach, C., Scalbert, A., Morand, C., Remesy, C., & Jimenez, L. (2004). Polyphenols: Food sources and bioavailability. American Journal of Clinical Nutrition, 79, 727–747. https://doi.org/10.1093/ajcn/79.5.727

Martínez-Villaluenga, C., Frias, J., Vidal-Valverde, C., (2008). Alpha-galactosides: antinutritional factors or functional ingredients? Critical Reviews in Food Science and Nutrition 48, 301–316.

McKnight, J.R., Satterfield, M.C., Jobgen, W.S., Smith, S.B., Spencer, T.E., Meininger, C.J., McNeal, C.J., Wu, G., (2010). Beneficial effects of l-arginine on reducing obesity, potential mechanisms and important implications for human health. Amino Acids 39, 349–357.

Messina, V., 2014. Nutritional and health benefits of dried beans. *American Journal of Clinical Nutrition* 100, 437S–442S.

Mishra SS, Ray RC, Panda SK, Montet D (2017) Technological innovations in processing of fermented foods. In: Ray RC, Montet D (eds) Fermented food part II: technological interventions. CRC Press, Boca Raton, pp 21–45

Mohite, B. V., Chaudhari, G. A., Ingale, H. S., & Mahajan, V. N. (2013). Effect of fermentation and processing on in vitro mineral estimation of selected fermented foods. *International Food Research Journal*, 20, 1373–1377.

Moktan, B., Roy, A., Sarkar, P.K., (2011). Antioxidant activities of cereal-legume mixed batters as influenced by process parameters during preparation of dhokla and idli, traditional steamed pancakes. *International Journal of Food Sciences and Nutrition* 62, 360–369.

Mugula, J. K. - Narvhus, J. A. Sorhaug, N. T. (2003) Use of starter cultures of lactic acid bacteria and yeasts in the preparation of togwa, a Tanzanian fermented food. International Journal of Food Microbiology, 83, 307-318.

Murdock, F. A., & Fields, M. L. (1984). B-vitamin content of natural lactic acid fermented cornmeal. *Journal of Food Science*, 49, 373–375.

Nalle, C.L., Ravindran, G., Ravindran, V., (2010). Influence of dehulling on the apparent metabolis¬able energy and ileal amino acid digestibility of grain legumes for broilers. *Journal of the Science of Food and Agriculture* 90, 1227–1231.

Nche, P. F. - Odamtten, G. T. - Nout, M. J. T.Rombouts, F. M. (1994). Dry milling and accelerated fermentation of maize for industrial production of kenkey a Ghanaian cereal food. Journal of Cereal Science, 20, 291-298.

Niveditha, V.R., Krishna Venkatramana, D., Sridhar, K.R., 2012. Fatty acid composition of cooked and fermented beans of the wild legumes (Canavalia) of coastal sand dunes. International Food Research Journal 19, 1401–1407.

Niveditha, V.R., Krishna Venkatramana, D., Sridhar, K.R., (2013). Cytotoxic effects of methanol extract of raw, cooked and fermented split beans of Canavalia on cancer cell lines MCF-7 and HT-29. *IIOAB Journal* 4, 20–23.

Nout, M. J. R. - Ngoddy, P.O. (1997). Technological aspects of preparing affordable fermented complementary foods. *Food Control*, 8, pp. 279-287.

Oh MR, Park SH, Kim SY, Back HI, Kim MG, Jeon JY, Ha KC, Na WT, Cha YS, Park BH, Park TS, Chae SW (2014) Postprandial glucose-lowering effects of fermented red ginseng in subjects with impaired fasting glucose or type 2 diabetes: a randomized, double-blind, placebo-controlled clinical trial. BMC Comp Alt Med 14. https://doi.org/10.1186/1472-6882-14-237

Omemu, A. M. (2011). Fermentation dynamics during production of ogi, a Nigerian fermented cereal porridge. Report and Opinion, 3, 8–17

Ortiz, D., Nkhata, S., Buechler, A., Rocheford, T., & Ferruzzi, M. G. (2017). Nutritional changes during biofortified maize fermentation (steeping) for ogi production. *The FASEB Journal*, 31, 1.

Osman, M. A. (2011). Effect of traditional fermentation process on the nutrient and antinutrient contents of pearl millet during preparation of Lohoh. *Journal of the Saudi Society of Agricultural Sciences*, 10, 1–6.

Parvez, S., Malik, K.A., Ah Kang, S., Kim, H.Y., (2006). Probiotics and their fermented food products are beneficial for health. *Journal of Applied Microbiology* 100, 1171–1185.

Phorbee, O. O., Olayiwola, I. O., & Sanni, S. A. (2013). Bioavailability of beta carotene in traditional fermented, roasted granules, gari from bio-fortified cassava roots. *Food and Nutrition Sciences*, 4,1247–1254.

Pranoto, Y., Anggrahini, S., & Efendi, Z. (2013). Effect of natural and Lactobacillus plantarum fermentation on *in vitro* protein and starch digestibilities of sorghum flours. Food Bioscience, 2, 46–52. https:// doi.org/10.1016/j.fbio.2013.04.001

Prinyawiwatkul, W., Beuchat, L.R., McWatters, K.H., Phillips, R.D., (1996). Changes in fatty acid, simple sugar, and oligosaccharide content of cowpea (Vigna unguiculata) flour as a result of soaking, boiling, and fermentation with Rhizopus microsporus var. oligosporus. *Food Chemistry* 57, 405–413.

Randhir, R., Shetty, K., (2007). Mung beans processed by solid-state bioconversion improves phe¬nolic content and functionality relevant for diabetes and ulcer management. *Innovative Food Science and Emerging Technologies* 8, 197–204.

Reale, A., Konietzny, U., Coppola, R., Sorrentino, E., & Greiner, R. (2007). The importance of lactic acid bacteria for phytate degradation during cereal dough fermentation. *Journal of Agricultural and Food Chemistry*, 55, 2993–2997. https://doi.org/10.1021/jf063507n

Rebello, C., Greenway, F.L., Dhurandhar, N.V., (2014). Functional foods to promote weight loss and satiety. Current Opinion in Clinical Nutrition and Metabolic Care 17, 596–604.

Reyes-Bastidas, M., Reyes-Fernández, E.Z., López-Cervantes, J., Milán-Carrillo, J., Loarca-Piña, G.F., Reyes-Moreno, C., 2010. Physicochemical, nutritional and antioxidant properties of tem¬peh flour from common bean (*Phaseolus vulgaris* L.). *Food Science and Technology International* 16, 427–434.

Reyes-Bastidas, M., Reyes-Fernández, E.Z., López-Cervantes, J., Milán-Carrillo, J., Loarca-Piña, G.F., Reyes-Moreno, C., (2010). Physicochemical, nutritional and antioxidant properties of tempeh flour from common bean (*Phaseolus vulgaris* L.). *Food Science and Technology International* 16, 427–434.

Reyes-Moreno, C., Cuevas-Rodríguez, E.O., Milán-Carrillo, J., Cárdenas-Valenzuela, O.G., Barrón- Hoyos, J., (2004). Solid state fermentation process for producing chickpea (Cicer arietinum L) tempeh flour. Physicochemical and nutritional characteristics of the product. *Journal of the Science of Food and Agriculture* 84, 271–278.

Riat, P., Sadana, B., (2009). Effect of fermentation on amino acid composition of cereal and pulse based foods. *Journal of Food Science and Technology* 46, 247–250.

Rui, X., Wen, D., Li, W., Chen, X., Jiang, M., Dong, M., (2015). Enrichment of ACE inhibitory peptides in navy bean (*Phaseolus vulgaris*) using lactic acid bacteria. *Food and Function* 6, 622–629.

Santoyo, M. C. Loiseau, G. Rodriguez Sanoja, R.Guyot, J. P. (2003) Study of starch fermentation at low pH by Lactobacillus fermentum Ogi E1 reveals uncoupling between growth and α-amylase production at pH 4,0. International Journal of Food Microbiology,80, pp. 77-87.

Sarasa, S. and Nath, N., (1985). Studies on the acceptability and palatability of certain new idli like products, *J. Food Sci. Technol.*, 22, 167.

Satish KR, Kanmani P, Yuvaraj N, Paari KA, Pattukumar V, Arul V. Traditional Indian fermented foods: A rich source of lactic acid bacteria. International Journal of Food Science and Nutrition. 2013;64:415-428. DOI: 10.3109/09637486.2012.746288

Sharma, G., Srivastava, A.K., Prakash, D., (2011). Phytochemicals of nutraceutical importance: their role in health and diseases. Pharmacology 2, 408–427.

Shimelis, E.A., Rakshit, S.K., 2008. Influence of natural and controlled fermentations on α-galactosides, antinutrients and protein digestibility of beans (*Phaseolus vulgaris* L.). *International Journal of Food Science and Technology* 43, 658–665.

Shimelis, E.A., Rakshit, S.K., (2008). Influence of natural and controlled fermentations on α-galactosides, antinutrients and protein digestibility of beans (*Phaseolus vulgaris* L.). *International Journal of Food Science and Technology* 43, 658–665.

Shimelis, E.A., Rakshit, S.K., (2008). Influence of natural and controlled fermentations on α-galactosides, antinutrients and protein digestibility of beans (*Phaseolus vulgaris* L.). *International Journal of Food Science and Technology* 43, 658–665.

Sindhu, S. C. - Khetarpaul, N. (2001). Probiotic fermentation of indigenous food mixture: Effect on antinutrients and digestibility of starch and protein. *Journal of Food Composition and Analysis,* 14, 2001,pp. 601-609.

Smith, A.P., (2012). Legumes and well-being in the elderly, a preliminary study. *Journal of Food Research* 1, 165–168.

Starzyńska-Janiszewska, A., Stodolak, B., (2011). Effect of inoculated lactic acid fermentation on antinutritional and antiradical properties of grass pea (*Lathyrus sativus* 'Krab') flour. Polish *Journal of Food and Nutrition Sciences* 61, 245–249.

Starzyńska-Janiszewska, A., Stodolak, B., Mickowska, B., (2014). Effect of controlled lactic acid fer¬mentation on selected bioactive and nutritional parameters of tempeh obtained from unhulled common bean (*Phaseolus vulgaris*) seeds. *Journal of the Science of Food and Agriculture* 94, 359–366.

Starzyńska-Janiszewska, A., Stodolak, B., Wikiera, A., (2015). Proteolysis in tempeh-type products obtained with Rhizopus and Aspergillus strains from grass pea (Lathyrus Sativus) seeds. *Acta Scientiarum Polonorum. Technologia Alimentaria* 14, 125–132.

Steinkraus, K.H. (1996) Handbook of Indigenous Fermented Foods, Second edition. Marcel Dekker, New York

Stewart GG (2016) Saccharomyces species in the production of beer. Beverages 2(4):34. https://doi.org/10.3390/beverages2040034

Sugano, M., (2005). Soy in Health and Disease Prevention. CRC Press, Boca Raton, FL.

Suwanmanon, K., Hsieh, P.C., 2014. Effect of γ-aminobutyric acid and nattokinase-enriched fer¬mented beans on the blood pressure of spontaneously hypertensive and normotensive Wistar- Kyoto rats. *Journal of Food and Drug Analysis* 22, 485–491.

Svanberg, U., & Lorri, W. (1997). Fermentation and nutrient availability. *Food Control,* 8(5/6), 319–327.

Swain MR, Anandharaj M, Ray RC, Parveen RR (2014) Fermented fruits and vegetables of Asia: a potential source of probiotics. Biotechnol Res Int. https://doi.org/10.1155/2014/250424

Torino, M.I., Limon, R.I., Martinez-Villaluenga, C., Makinen, S., Pihlanto, A., Vidal-Valverde, C., Frias, J., (2013). Antioxidant and antihypertensive properties of liquid and solid state fermented lentils. *Food Chemistry* 136, 1030–1037.

Torres, A., Frias, J., Granito, M., Vidal-Valverde, C., (2006). Fermented pigeon pea (Cajanus cajan) ingredients in pasta products. *Journal of Agricultural and Food Chemistry* 54, 6685–6691.

Towo, E., Matuschek, E., & Svanberg, U. (2006). Fermentation and enzyme treatment of tannin sorghum gruels: Effects on phenolic compounds, phytate and in vitro accessible iron. *Food Chemistry,* 94, 369–376. https://doi.org/10.1016/j.foodchem.2004.11.027

Untersmayr, E., & Jensen-Jarolim, E. (2008). The role of protein digestibility and antacids on food allergy outcomes. *Journal of Allergy and Clinical Immunology,* 121, 1301–1310. https://doi.org/10.1016/j.jaci.2008.04.025

Vaishali, V. A., Medha, K. G., & Shashi, A. C. (1997). Effect of natural fermentation on in vitro bioavailability in cereal-legume mixtures. *International Journal of Food Science and Technology,* 32, 29–32.

Walker GM, Hill AE (2016) Saccharomyces cerevisiae in the production of whisk(e)y. *Beverages* 2(4):38. https://doi.org/10.3390/beverages2040038

Wu, H., Rui, X., Li, W., Chen, X., Jiang, M., Dong, M., 2015. Mung bean (Vigna radiata) as probi¬otic food through fermentation with Lactobacillus plantarum B1-6. *LWT - Food Science and Technology* 63, 445–451.

Yadav, M., Singh, P., Kaur, R., Gupta, R., Gangoliya, S.S., Singh, N.K., (2013). Impact of food phytic acid on nutritions, health and environment. *Plant Archives* 13, 605–611.

Yeap, S.K., Beh, B.K., Ali, N.M., Mohd Yusof, H., Ho, W.Y., Koh, S.P., Alitheen, N.B., Long, K., (2014). *In vivo* antistress and antioxidant effects of fermented and germinated mung bean. BioMed Research International: 694842.

Yeap, S.K., Mohd Ali, N., Mohd Yusof, H., Alitheen, N.B., Beh, B.K., Ho, W.Y., Koh, S.P., Long, K., (2012). Antihyperglycemic effects of fermented and nonfermented mung bean extracts on alloxan-induced-diabetic mice. *Journal of Biomedicine and Biotechnology*:285430.

Yeap, S.K., Mohd Yusof, H., Mohamad, N.E., Beh, B.K., Ho, W.Y., Ali, N.M., Alitheen, N.B., Koh, S.P., Long, K., (2013). *In vivo* immunomodulation and lipid peroxidation activities contributed to chemoprevention effects of fermented mung bean against breast cancer. Evidence-Based Complementary and Alternative Medicine 2013:708464.

Yerlikaya O (2014) Starter cultures used in probiotic dairy product preparation and popular probiotic dairy drinks. *Food Sci Tech* (Campinas) 34(2):221–229

Yousif, N. E. - El Tinay, A. H. (2000). Effect of fermentation on protein fractions and in vitro protein digestibility of maize. *Food Chemistry,* 70, 2000, pp. 181-184.

Zhang, G., Xu, Z., Gao, Y., Huang, X., & Yang, T. (2015). Effects of germination on the nutritional properties, phenolic profiles, and antioxidant activities of buckwheat. *Journal of Food Science,* 80, H1111–H1119. https://doi.org/10.1111/1750-3841.12830

9

Oxidative Changes in Processed Foods

Dev Kumar Yadav, Chaitra G.H., Gopal Kumar Sharma and A.D. Semwal

Grain Science and Technology Division, DRDO- Defence Food Research Laboratory, Mysore, Karnataka

Introduction

Lipids are among the major components of food of plant and animal origin (food lipid). There is no precise definition available for the term lipid; however, it always includes a broad category of compounds that have some common properties and compositional similarities. Lipids are materials that are sparingly soluble or insoluble in water, but soluble in selected organic solvents like benzene, chloroform, ether, hexane, and methanol. Together with carbohydrates and proteins, lipids constitute the principal structural components of tissues (Erwin Wsowicz et al., 2004). Lipid oxidation is one of the major reasons that foods deteriorate and is caused by the reaction of fats and oils with molecular oxygen resulting in off-flavours that are generally called rancidity. Exposure to light, pro-oxidants and elevated temperature will accelerate the reaction (Velasco, J et al., 2010). Rancidity is related to characteristic off-flavour and odour of the oil. There are two major causes of rancidity. One occurs when oil reacts with oxygen and is called oxidative rancidity (Ribeiro et. al., 1993). The other cause of rancidity is by a combination of enzymes and moisture. Enzymes like lipases liberate fatty acids from the triglyceride to make di and monoglycerides and free fatty acids and such liberation of free fatty acids is named hydrolysis. Hydrolysis is additionally caused by chemical process that's prompted by factors like heat or presence of water. Rancidity caused by hydrolysis is called hydrolytic rancidity. Oxidation cares mainly with the unsaturated fatty acids. Oxidative rancidity is off interest because it results in the event of unfavourable off-flavours which will be detected early within the development of rancidity, more so than in the case of hydrolytic rancidity. Oxidation can alter the flavour and nutritional quality of foods and produce toxic compounds, all of which may make the foods less acceptable or unacceptable to consumers. Oxidation products typically include low relative molecular mass compounds that are volatile also as undesirable off-flavour compounds (Food lipids).

The most common and important process by which unsaturated fatty acids and oxygen interact is a free radical mechanism characterized by three main phases:

Primary oxidation

The products which are formed by the reaction of an alkyl radical, which is formed due to the reaction to light or heat (initiation), in the presence of oxygen to form peroxy free radical. The formed peroxy free radical interacts with an un-attacked unsaturated fatty acid to form a alkyl free radical and fat hydroperoxide (propagation). This resulted product is odourless and tasteless. The reaction continues till depletion of oxygen occurs or when two unstable radicals react or when a fatty radical reacts with a stable antioxidant radical (termination). This mechanism includes three steps, namely initiation, propagation and termination, which known as auto-oxidation (Labuza., 1971 & Frankel., 2005).

		catalyst	
Initiation:	RH + O_2	→	R* + *OOH
		catalyst	
	RH	→	R* + *H
Propagation	R* + O_2	→	ROO*
	ROO* + RH	→	ROOH + R*
Termination	R* + R*	→	RR
	ROO* + R*	→	ROO*

Initiation

The direct reaction between lipid molecules and oxygen molecule is very improbable because lipid molecule is during a singlet electronic state and therefore the oxygen molecule found in triplet state. to bypass this spin restriction, oxygen are often activated by any of the subsequent three initiation mechanisms: (1) formation of singlet oxygen; (2) formation of partially reduced or activated oxygen species like peroxide, superoxide , or hydroxyl radical; and/or (3) formation of active oxygen–iron complexes (ferryl iron or ferric–oxygen– ferrous complex). Additionally, the oxidation of fatty acids may occur directly or indirectly through the action of enzyme systems, of which three major groups are involved: microsomal enzymes, peroxidases, and dioxygenases, like lipoxygenase or cyclooxygenase (Frankel., 2005 & Nawar, W. W., 1996).

Propagation

Propagation reaction involves the chain reaction process and in general includes the following:

R
Lipid radical
O_2
Propagation
R R R
OOH H + OO^-
Lipid peroxide Lipid peroxyl radical

Termination

Two types of termination reactions are involved in the process of breaking of repeating sequence of propagating steps, they are radical–radical coupling and radical– radical disproportionation a process during which two stable products are formed from A• and B• by an atom or group transfer process. In both cases, non-radical products are formed. Therefore, the termination reactions are not always proficient. When coupling gives rise to tertiary tetroxides, they decompose to peroxyl radicals at temperatures above 80°C and to alkoxyl radicals at temperatures above 30°C. Secondary and first peroxyl radicals, on the opposite hand, terminate efficiently by a mechanism during which the tetroxide decomposes to offer molecular oxygen, an alcohol, and a carbonyl compound (Hoffman, 1989).

Secondary oxidation

Products are formed when the hydroperoxides decompose to secondary oxidation products as a result of heating, radiation, or the presence of heavy metals (Cu, Fe) and other radical initiating agents. The secondary oxidation products are formed by either peroxide scission alone or simultaneous peroxide and chain scission. Chain scission leads to short-chain volatiles such as aldehydes, ketones, alcohols and acids, which cause the characteristic off-flavours and odours of rancid fats and oils. The addition of antioxidants can retard autoxidation, as they're radical scavengers and by interrupting the chain reaction prevent or hamper the propagation of oxidation. Synthetic antioxidants used are phenols such as BHA, BHT, TBHQ and propyl gallate Tocopherols (Vitamin E) naturally present in oil act as natural antioxidants (Hoffman, 1989)

Factors influencing oxidative stability

It is important to remember of the factors that influence oxidative stability to make sure the longest shelf-life possible for oil.

Antioxidants

Antioxidants retard the onset of natural action, thereby extending the shelf-life of fats and oils and nutrient, however cannot stop it. It's an equivalent for artificial antioxidants like BHA, BHT, TBHQ and natural antioxidants like tocopherols. Antioxidants will act either as primary chain breaking antioxidants, or as secondary preventative antioxidants. Most of the common food antioxidants (AH) act as chain breakers by donating matter atoms to the chemical compound radicals intentional throughout initiation, thereby halting or retardation down the propagation of natural action (Hamilton, 1994).

A* + ROO*

Non-radical products

A* + A*

ROO* + AH ROOH + A*

The free radical A* does not participate in propagation steps as it is stabilised by resonance

Secondary antioxidants reduce the rate of chain initiation by various mechanisms such as scavenging oxygen, decompose hydroperoxides to non-radical species, binding to metal ions, absorb ultraviolet (UV) radiation or deactivate singlet oxygen (Hamilton, 1994).

Pro-oxidants: Pro-oxidants in oil have a detrimental effect on oil stability. Metals act as pro-oxidants by electron transfer whereby they liberate radicals from fatty acids or hydroperoxides as in the following reactions

$M^{(n+1)+}$ + RH M^{n+} + H+ R*

ROOH + M^{n+} RO* + OH- +$M^{(n+1)+}$

ROOH + $M^{(n+1)+}$ ROO* + H+ +M^{n+}

Two of the more active metals to induce oxidation are copper and iron of which copper is the most pro-oxidative (Bondet et al., 2000).

Oxygen availability

The availability of element is a vital rate-determining issue as oxidisation cannot crop up while not element (Berger, 1994). the speed of lipid oxidisation measured by hexanal formation magnified with increasing concentrations of element (1.2 %, 4.5 %, 10.0 % and 15.4 %) in a very closed system. it's conjointly accepted that samples with a high area in grips with air oxidize quicker (Gordon, Mursi and Rossell,

1994). this can be clearly illustrated in a very study with extracted crude oil hold on beneath 3totally different storage conditions; in a very capped flask, open flask and capped flask beneath gas atmosphere. There was very little distinction within the oxidisation rate between the open and capped flask, that indicates thatoxidisation rate depends on the relation between oil area exposed to air and sample volume, whereas the capped flask beneath gas showedlittle aerobic activity. element is replaced by utilizing a protecting gas observe like gas blanketing which will defend oil in storage tanks, throughout bulk transport and once pre-packed against oxidisation (O'Brien, 1998).

Light

Light can accelerate auto-oxidation by favouring the formation of free radicals in the initiation step, acting as a catalyst for hydrogen abstraction and in the decomposition of hydroperoxides. This effect is different from photo-oxidation which takes place in the presence of photo sensitizer. The oxidation process is likely to be accelerated during retail storage as the kernels are often held at ambient temperature and relative humidity and are also exposed to light that catalyses oxidation (Sattar *et. al.*, 1989). Here the multifactorial effect on the oxidative stability of the nut is clearly evident. Light, depending on its intensity, wavelength, duration of exposure, absorption by the product, presence of sensitizers, temperature and the amount of available oxygen may induce oxidation in nuts and nut oils (Özdemir and the Devres,1999b).

Temperature

An increase in temperature ends up in a significant reduction within the length of the induction period. The speed of oxidation increases exponentially with temperature. Additionally, there's a powerful interaction between temperature and oxygen because the oxygen solubility decreases as temperature increases (Dobarganes et. al., 1999). At temperatures larger than 130–140 °C the hydroperoxides are unstable anddecompose at a rate larger than that of their formation. Thus, the secondary compounds of oxidation constitute the foremost significant products in processes that, just like the frying of foods, are administrated at high temperatures (Masson et al., 1997; Stevenson et al., 1984)

Fatty acid composition

The degree of unsaturation is one in every of the foremost determining factors within the speed of lipid oxidation. The relative autoxidation rate of oleic, linoleic and linolenic acids has been reported to be 1:40–50:100 as detected by oxygen absorption, and 1:12:25 as detected by hydroperoxide formation (Holman and Elmer, 1947). In mixtures, because it occurs in nature, the oxidation rate of the foremost unsaturated acid is determinant and differences are much under that expected from the results obtained in pure lipids oxidized separately (Bolland, 1949). Additionally, it's been proven that the oxidation of saturated fatty acids is extremely low, remaining unaltered even when the degradation of the unsaturated fatty acids is elevated. Nevertheless, changes of saturated fatty acids are significant at elevated temperatures (Swern, 1961). It's well established that the speed of oxidation also depends on the position and geometrical configuration of double bonds, and also the length of the acid chain (Sahasrabudhe and Farne, 1964).

Oxygen content

Several studies have unambiguously demonstrated that storage at high oxygen concentrations results in more pronounced lipid oxidation than storage under low oxygen concentrations (Ribeiro et al., 1993). The absolute absence of oxygen would prevent the oxidative rancidity of lipids, as oxygen is important to propagate the reaction (Velasco, 2010).

Effect of processing method on rancidity

Roasting

In hazelnut oleic and saturated fatty acids increased, while linoleic aciddecreased with increase in roasting temperature and time. Similarly, a rise of triacylglycerols containing monounsaturated fatty acid and a decrease of this containing linolic acid were found within the roasted samples. Roasting caused a modest decrease of the beneficial phytosterols (maximum 14.4%) and vitamin E homologues (maximum 10.0%) and a negligible increase of the Trans fatty acids (Amaral et al. 2006). In peanuts ESR spin trapping technique could determine radical generating reactions in the very early stage of fat deterioration. Oxidative reactions reduced the time period of peanuts also as their sensory quality. Roasting has a controversial influence on the stability of peanuts. With increasing roasting temperature and time the oxidative stability of peanuts was improved and shelf life prolonged, possibly due to the formation of antioxidant Maillard reaction products. A correlation was found between the amount of deoxyosones as reactive Maillard reaction intermediates and shelf life of roasted nuts (Cämmerer and Kroh, 2009). In Desmayo Largueta variety of almond Rancidity increased with the treatment time up to a maximum value and over-roasting produced antioxidant products due to Maillard reaction. Over- roasting also decreased sweetness and increased bitterness and grittiness. The variety of almond can play a crucial role within the time period of the merchandise. The content of tocopherol, lipids and peroxide values depended on the variety as well as the soil and climate conditions where the nuts are grown (Gou et al. 2000).

Blanching

Blanching at 100°C greatly increased the rate of peroxide formation. Lipase appeared to be present and active. Peroxide and Kreis values increased many times faster in oil stored as such than in the meats (Musco and Cruess, 1954). Blanching of hazelnut kernels in the production of hazelnut meal was found to decrease the levels of free fatty acids that could reduce hydrolytic rancidity during storage. However, oxidation occurred in the blanched hazelnut meal that could lead to oxidative rancidity. The concentration of rancid compounds should also be measured to determine the effect of blanching on rancidity development in hazelnut meal. Native enzymes in hazelnuts should be inactivated to improve the stability of hazelnut meal during storage. A method that could limit oxidation during the treatment should be chosen to inactivate the enzymes and to reduce oxidation during storage (Özdemir et al. 2001).

Ageing

In Macadamia nut Oil High- resolution electrospray ionization Fourier transform ion cyclotron resonance mass spectrometry was used. Mass analysis of aged macadamia oil revealed that oils obtained by the cold press method are more vulnerable toageing than those obtained using modified Soxhlet or accelerated solvent extraction methods; suggesting that the activity of enzymes must be minimized, especially in traditional cold pressed extraction. Also, indicates that enzymatic rancidity may play a bigger role than radical oxidation (Proschogo et al.,2012).

Characteristic changes during rancidity sensorial changes

Sensory evaluation is generally considered to be the most reliable indicator of rancidity and measurement of flavor quality of plant oils (Wamer and Frankel, 1985). In a study on canola oil stored and characterized for consumer acceptability a characteristic of rancidity was the detection of a painty odour and taste (Malcolmson et aI, 1996). This characteristic flavor is generally accepted as detection of rancidity. Along with the degradation of the unsaturated fatty acids present that contributes to the off-flavors and odors, some components of the unsaponifiable also matters in detection of rancidity. Flavor deterioration is the most common concern regarding the use of rancid fats and oils (Haumann, 1993). Rancidity in fats and oils has a characteristic, unpalatable off-flavor and odour in oils, which can be picked up easily by subjective sensory appraisal (Hamilton, 1994). Secondary oxidation products like short-chain aldehydes cause the quality off-flavor, which relying on their structure and thus the amounts formed, because odors like beany, grassy, painty, fishy, tallowy or plain rancidity (Hoffman, 1989). Malcolmson (1995) further states that sensory evaluation is the application of knowledge and skills of various scientific disciplines among them food science, psychology, physiology, mathematics and statistics. Sensory evaluation of oils is limited mainly to the senses of taste and smell (Warner, 1995).

Nutritional changes

The oxidized fats are considered as rendered unpalatable because of the deterioration in flavour and appearance long before the changes have appreciably reducednutritive value or created toxicity (Gurr, 1988). The most susceptible to oxidation are those oils which are rich in polyunsaturated fatty acids. The nutritional effect of oxidation is to reduce the essential fatty acid content of edible fats. But the significance of that occurrence for nutrition is quite low since losses are usually small in relation to the total content of polyunsaturated fatty acids supplied by these susceptible oils. More serious nutritional problem of lipid oxidation is affected by interactions of lipid oxidation products with other food components, mainly with vitamins and proteins (Sanders, 1989).

Physical change viscosity

According to Kim, J. et al., (2010), viscosity decreases with an increase in unsaturation and increases with high saturation and polymerization. Viscosity also depends on sheer stress and temperature. When the temperature increases, the kinetic energy also increases which helps in enhancement of movement of the molecules and decreases the intermolecular forces. The layers of the liquid easily skip each other and thus contribute to the reduction of viscosity. This phenomenon has been reported by other workers since oil viscosity depends on molecular structure and reduces with the unsaturation of fatty acids (Kim, J et al., 2010).

Refractive index

There is increase in the index of refraction in deep fried or re used oil due to the increase in the levels conjugated fatty acids as a result of thermal degradation during the frying process. The refractive index during heating of oil increases as more conjugated acids are formed. Refractive index varies with temperature and wavelength (Chakrabarty, 2003).

Density

Densities of fatty acids and glycerides increases with increase in their unsaturation and decrease in molecular weight. Fats and their derivatives have higher density in their solid form than in liquid state. When they are subjected to solidification or melting, they show shrinkage or expansion much greater than the thermal expansion of solid or liquid phase. Density relationships for mixtures of mixed triglycerides are more complex than the density of pure single acid glyceride (Chakrabarty, 2003).

Melting point

Melting point of fatty materials may be a vital property as freezing point identifies the fabric, and it's also used for several technological applications. Melting points of fatty acids increase with increasing chain length and reduce with unsaturation. Pure fatty acids, esters and glycerides often exhibit distinct melting points as they are polymorphic. Even and odd numbered carbon fatty acids and glycerides show a regular alteration of melting points (Chakrabarty, 2003)

Health effects

While rancid oil may taste bad, it doesn't normally cause you to sick, a minimum of not within the short term. Rancid oil does contain free radicals which may increase your risk of developing diseases like cancer or heart condition down the road. Rancid oils may produce damaging chemicals and substances which will not cause you to immediately ill, but can cause harm over time. Chemicals like peroxides and aldehydes can damage cells and contribute to atherosclerosis. Free radicals produced by rancid oil also can damage DNA in cells. Produced by toxins also as by normal bodily processes, free radicals can cause damage to arteries also act as carcinogens, substances which will cause cancer. If oxidative rancidity is present in severe quantities, a possible hazard may exist. Higher levels of malonaldehyde are found in rancid foods. Malonaldehyde may be a decomposition product of polyunsaturated fatty acids. This chemical has been reported to be carcinogenic and a possible hazard does exist. Eating morerancid oil will leads to accelerated aging, raised cholesterol levels, obesity and weight gain. Daily consumption increases the risk of degenerating diseases such as cancer; diabetes; Alzheimer's disease; and atherosclerosis, a condition in which artery walls thicken due to a build-up of fattymaterials.

According to a study from the University of Basque Country, the breakdown rate and total formation of toxic compounds depends on the sort of oil and temperature. Initially, the oil decomposes into hydroperoxides, then into aldehydes (Susan Okparanta et al., 2018). Claims of nutritional implications of the consumption of oxidised fats and oils are varied. According to Sanders (1994) the symptoms of rancid fat toxicity are diarrhoea, poor growth rate, myopathy (replacement of healthy muscle with scar tissue), hepatomegaly (enlarged liver), steatites or yellow fat disease, hemolytic anemia and secondary deficiencies of vitamins A and E. Evidence shows that dietary oxidation products may cause arterial injury, artherosclerotic plaque formation and thrombosis/ spasm which are potentially dangerous (Haumann, 1993).

Consumption of lipid oxidation merchandise ends in liver and kidney weights, cellular damage in numerous organs, altered fatty acid composition of tissue lipids, cardiac fibrotic lesions, and hepatic bile duct lesions (Sanders, 1989; Addis & Warner, 1991; Kubow, 1990; Eder, 1999). Studies at the viable pathological significance of lipid oxidation products were involved on the impact of lipid peroxides, secondary products of lipid oxidation, mainly malondialdehyde and LDL cholesterol oxidation products (Addis, 1986). The primary merchandise of autoxidation – carboxylic acid peroxides - are probably no longer effectively absorbed from the gut, and therefore themaximum

mighty deleterious consequences of lipid peroxides appear at the gastrointestinal mucosa. But it is mentioned the probability, that oxidation in vivo and weight loss program are each the assets of serum lipid peroxides (Addis, 1986). Fatty acid peroxides are shown to boost up all 3 levels of atherosclerosis: initiation – endothelial injury, progression –accumulation of plaque, and termination – thrombosis (Kubow, 1990).

Dietary lipid peroxides participate inside the improvement of cancer in humans. It become demonstrated a sturdy reaction among lipid peroxides and DNA (Addis, 1986). Low molecular merchandise of decomposition of fatty acid peroxides, however, are absorbed into the circulatory machine and incorporated into the liver or have access to other body tissues. Malondialdehyde, a secondary made of lipid oxidation, has received a whole lot attention. Since it is a bifunctional aldehyde, it's a truly reactive compound in cross-linking reactions with DNA and proteins (Addis, 1986; Kubow, 1990). The toxicity of oxidized cholesterols has been demonstrated in numerous studies. The oxysterols are absorbed from the intestinal tract and are transported inside the blood to arterial deposition sites at rates nearly like ldl cholesterol (Kubow, 1990). There is great evidence that a few ldl cholesterol oxidation merchandise are effective atherogenic marketers in vivo and in vitro. They have additionally cytotoxic and mutagenic properties (Addis & Warner, 1991; Osada et al., 1998). In summary, it need to be found out that not a single product but a mixture of the above groups of lipid oxidation merchandise can arise in each day diets. For this purpose foods should be covered in any manner to reduce their concentration in ingredients and cast off their deleterious results. Thus the usage of natural antioxidants has been gaining widespread importance (Johnson, 2001; Virgili et al.,2001)

Subjective and objective methodologies to study oxidative changes free fatty acid

Free carboxylic acid is tertiary product of rancidity, which increased during storage. Free carboxylic acid may be a measure of hydrolytic rancidity (the extent of lipid hydrolysis by lipase action). The FFA determination measures the number of hydrolytic activity that has occurred within the oil. The share free fatty acids present within the oil are measured by the equivalent amount of hydrated oxide needed to neutralize the FFA (Sonntag, 1979b). Within the calculation of the FFA percentage, the belief is formed that the typical mass of the fatty acids is that of monounsaturated fatty acid (Sonntag, 1979b). The number of FFA present is of importance, not only because it indicates hydrolytic activity, but also because FFA includes a pro-oxidant effect, the intensity of which is expounded to FFA concentration (Frega, Mozzon and Lercker, 1999)

Peroxide value

Peroxide value is that the first measurement of oils rancidity and it gives us a concept of oils' freshness and storage conditions. The PV measures the hydroperoxides formed during the initial stages of oxidative rancidity of fats and oils (Hahm and Min, 1995; O'Brien, 1998). The tactic generally used relies on an iodometric titration with standardized chemical compound which measures the iodine liberated from iodide by the peroxides present within the oil (Rossell, 1994). The PV is expressed in terms of mili-equivalents of oxygen per 1 kg of fat (m.eq/kg). PV has shown good correlation with flavor scores but it's to be kept in mind that the PV is restricted to the initial stages of oxidation, because it reaches a peak value then oxidizes to secondary oxidation products (Hahm and Min, 1995; O'Brien, 1998). This implies that prime PVs usually give poor flavor scores. The peroxides value will increase during the first a part of the lifetime of oils and it will then decrease in additional advanced stages of oxidation when more oxidated substances are produced. These new substances are in command of colour and aroma changes associated with rancidity (S. Okparantaetal., 2018).

Anisidine value (AV)

The anisidine value measures the secondary products of oxidation. Aldehydes are one of the products of secondary oxidation and also they mainly contribute to off-odours. In the anisidine value procedure, the reaction of u- and p- aldehydes (primarily 2-alkenals) takes place with p- anisidine reagent in the presence of acetic acid (Robards et al, 1988; White, 1995). This results in the production of Schiff base compound that further leads to the formation of a yellowish colour that is measured at 350nm. If the aldehyde as a double bond conjugated to the carbonyl double bond, the molar absorbance will increase by a factor of four to five. Thus, anisidine value measures mainly two alkenals (White, 1995).

Thiobarbituric acid (TBA)

The thiobarbituric acid test is an efficient method that is used frequently to measure oxidation in foods, particularly in meats. This test is predicated on the very fact that the oxidation product is react with TBA to offer compounds that absorb light at 532-535 nm. The reaction is not specific and different oxidation compounds that respond to this test are known as TBA reactive substances or TBARS (Pokorný et al., 2005). When the condensation of two molecules takes place with a molecule of malonaldehyde (it is an oxidation compound in oils that is produced only from poly unsaturated fatty acids with three or more double bonds), coloured complex is formed. Also,hydroperoxides respond to this test because when thermal acidic conditions are applied, they decompose into various products that react with TBA. Urea, sugars, oxidised proteins or other oxidised components that are present in foods can react with TBA to form coloured components and, this is the reason that this test can be carried out on whole foods, and it may give information on oxidation damage to material other than the extractable fat itself (Rossell, 1994).

Conjugated diene and triene value

When polyunsaturated fatty acids are oxidized to form hydroperoxides, a shift in the position of the double bond occurs and they become conjugated. The extent of covalent bond displacement is directly associated with the degree of peroxidation that has occurred within the oil and thus the quantity of oxidation of the oil (Rossell, 1994; White, 1995). The conjugated diene value (CV) is expressed as a percentage of conjugated dienoic acid within the oil and is a sign of primary oxidation (White, 1995). The conjugated acids absorb ultra violet (UV) light with a maximum between 232 and 234 nm. CV value is a simple method whereby the oil is dissolved in a volumetric flask in a solvent such as iso-octane, read on a spectrophotometer at 232 nm against the solvent as blank and calculated in percentage (White, 1995). According to White (1995) the CV values of oxidised oils range between ° and

6 %, depending on the type of oil. The CV accumulates to a certain percentage in the oil and then plateaus as the dienes break down further to other oxidation products. Trienes are frequently formed at this stage. The secondary oxidation products, particularly di-ketones and conjugated trienes, can also be measured by UV absorbance at 268 nm, although it has to be kept in mind that the absorption of various compounds overlaps in this range (Rossell, 1994).

Technical approaches to prevent rancidity high temperature short time processing

Increases in lipolysis were substantial in pasteurized milk during storage. The magnitude of these increases depended on the source and treatment of samples.Some milk supplies required higher pasteurization temperatures than others to inhibit lipolysis adequately. Varying the homogenization pressure from 105 to 211 kg/cm 2 did not appear to alter thermal inactivation of lipase or susceptibility of fat to lipolysis. Processing at 76.7°C for 16 s should be adequate to protect most milk from lipolytic problems for 7 days after pasteurization, but higher temperatures may be necessary for longer shelf-life of susceptible raw milk supplies. (Shipe, W.F and. Senyk, G. F, 1981)

Active packaging

Active packaging is a modern strategy in preventing lipid oxidation. Different active substances with different mechanisms of action have been considered for imparting antioxidant activity to active packaging systems, including free radical scavengers, metal chelators, ultraviolet (UV) absorbers, oxygen scavengers etc. antioxidant agents have been incorporated into active packaging systems in different forms, mainly including independent sachet packages, adhesive-bonded labels, multilayer films etc. (Fang Tian et al., 2013).

Antioxidant packaging systems

Antioxidants in food products Because of the significantly harmful effect of lipid oxidation, a variety of synthetic and natural antioxidants have been added directly into food products to inhibit oxidative reactions and preserve food quality and nutrition. Lipid oxidation can be controlled by preventing the formation of lipid hydroperoxides and free radicals, or by scavenging the free radicals generated in food system (W Chaiyasit et al., 2007). Packaging has long been used to extend the shelf life of foods by providing an inert barrier to external conditions (S. Sacharow, 2006). Active packaging goes beyond the normal role of packaging by imparting specific, intentional functionality to the packaging system. Active packaging is often designed to increase time period, impart post-package processing, or improve food safety and quality. Antioxidant packaging includes antioxidant substances in food packaging systems to impart antioxidant activity (M. Ozdemir and J. Floros, 2004).

Conclusion

The oxidation of any product mostly affects their eating and keeping quality negatively. The most prominent changes related to oxidative rancidity takes place due to thermal abuse, exposure to oxygenated conditions, presence of certain metal such as iron etc. and exposure to light as well. This takes place in a sequential chain such as chain initiation, propagation and termination. The chemical markers for assessing the oxidative quality of fresh as well as processed food products are peroxide value, free fatty acid content, anisidine value, quantity of conjugated dienes and trienes. The rancid flavour of the product makes it unpalatable and affects its sensory acceptability negatively. This also limits the shelf life/keeping quality for longer storage at varying climatic conditions. The generation of free radicals and off flavours due to high temperature processing do have negative health impact on the consumers. Reports are distinguished in mentioning that the prolonged consumption of rancid food causes obesity, cancer, accelerated aging, raised cholesterol level, atherosclerosis, diabetes, Alzheimer's disease etc. smart packing interventions and addition of natural/ chemical antioxidants are two major approaches to reduce the extent of rancidity in various food products. In addition various novel food processing techniques such as high pressure processing, ohmic heating, high temperature short time (HTST) processing, active packaging etc. can also play an important role in managing the oxidative stability of food products and its market value.

References

Addis, P. B. & Warner, G. J. (1991). The potential health aspects of lipid oxidation products in food. *in*: *Free Radicals and Food Additives.*, 77–119.

Addis, P. B. (1986). Occurrence of lipid oxidation products in foods. *Food Chem. Toxic*, 24, 1021–1030.

Amaral, J. S.; Casal, S.; Seabra, R. M. & Oliveira, B. P.P. (2006) Effects of roasting on hazelnut lipids. *J. Agric. Food Chem.*, 54, 1315 - 1321.

Berger, K. G. & Hamilton, R. J. Lipids and oxygen: is rancidity avoidable in practice? In Developments in oils and fats. Springer, 1995, 192-203.

Bolland, J. L. (1949) 'Kinetics of olefi n oxidation', Quar Rev., **3**, 1–21

Bondet, V.; Cuvelier, M. E. & Berset, C. (2000). Behaviour of phenolic antioxidants in a partitioned medium: focus on linoleic acid peroxidation induced by iron/ascorbic acid system. *J. Am. Oil Chem. Soc.*, 2000, **77**, 813–818.

Cämmerer, B. & Kroh, L. W. (2009)Shelf-life of linseeds and peanuts in relation to roasting. *Food Sci.Tech.*, 42, 545 - 549.

Chakrabarty. (2003). Chemistry and technology of fats and oils. Allied publishers, 47– 57.

Dobarganes, M. C.; Màrquez-Ruiz, G. & Velasco, J. (1999). Determination of oxidized compounds and oligomers by chromatographic techniques. *Frying of Foods.*, 143–161.

Eder, K. (1999). The effect of a dietary oxidized oil on lipid metabolism in rats. Lipids, **34**, 717–725.

Fang Tian.; Eric, A. Decker.; Julie, M. & Goddard. (2013). Controlling lipid oxidation of food by active packaging technologies. *Food Funct.*, 4, 669.

Fitch Haumann, B. (1993). Lipid Oxidation. Inform-Champaign, 4, 800-800.

Frankel, E. N. Chemistry of auto oxidation: mechanism, products and flavor significance. Flavor chemistry of fats and oils. Illinois: AOCS Press, 1–38.

Frankel, E. N. (2005). Lipid oxidation. The Oily Press.

Frankel, E. N. (1991). Recent advances in lipid oxidation. *Journal of the Science of Food and Agriculture.*, 54(4), 495-511.

Frega, N.; Mozzon, M. & Lercker, G. (1999). Effects of free fatty acids on oxidative stability of vegetable oil. *Journal of the American Oil Chemists' Society,* 76(3), 325-329.

Gordon, M. H. (1990). The mechanism of antioxidant action in-vitro," in Food Antioxidants. Elsevier Applied Science, 1990, 1–18.

Gordon, M. H.; Mursi, E. & Rossell, J. B. (1994). Assessment of thin-film oxidation with ultraviolet irradiation for predicting the oxidative stability of edible oils. *Journal of the American Oil Chemists' Society.*, 71(12), 1309-1313.

Gou , P.; Diýaz, I.; Guerrero, L.; Valero, A. & Arnau, J. (2000). Physicochemical and sensory property changes in almonds of Desmayo Largueta variety during roasting. *Food Sci. Technol. Int.*, 6, 1 – 7.

Gurr, M. I. Lipids: (1998). Products of industrial hydrogenation, oxidation, and heating. *in*: Nutritional and Toxicological Aspects of Food Processing. Taylor & Francis, London, 139–155.

Hahm T. S and Min D. B. (1995). Analyses of peroxide values and headspace oxygen, Methods to assess quality and stability of oils and fat-containing foods. AOCS Press, Champain, 146-158.

Hamilton. R. J. (1994). The chemistry of rancidity in foods. Rancidity in foods, Blackie Academic & Professional, 1-21.

Hoffmann, G. (1989). The chemistry of edible fats. In The chemistry and technology of edible oils and fats and their high fat products, 1–28.

Holman, R. T & Elmer. (1947). The rates of oxidation of unsaturated fatty acids and esters'. *J Am Oil Chem Soc.*, 24, 127–129.

Johnson, I. T. (2001). Antioxidants and antitumor properties. *in*: Antioxidants in Food. Wood head Publ. Ltd, Cambridge, 100–123.

Kim, J.; Kim, D. N.; Lee, S. H.; Yoo, S. H. & Lee, S. (2010). Correlation of fatty acid composition of vegetable oils with rheological behaviour and oil uptake. *Food Chem.*, 118, 398–402.

Kubow, S. Toxicity of dietary lipid peroxidation products. *Trends Food Sci. Technol.*, 1990, 1, 67–71.

Labuza, T. P. (1971). Kinetics of lipid oxidation in foods. CRC Critical Reviews in Food Science and Technology, **2**, 355–405.

Malcomson, L. J. (1995). Organization of a Sensory Evaluation Program. Methods to Assess Quality and Stability of Oils and Fats Containing Foods. American Oil Chemists' Society, 3, 37-49.

Masson, L.; Rorbert, P.; Romero , N.; Izaurieta, M. & Valenzuela, S. (1997). Performance of polyunsaturated oils during frying of potatoes in fast food shops: Formation of new compounds and correlation between analytical methods . *Grasas y Aceites.*, 48, 577 – 583.

Musco, D. D. & Cruess, W. V. (1954). Food Rancidity, Studies on Deterioration of Walnut Meats. *Journal of Agricultural and Food Chemistry.*, 2(10), 520-523.

Nawar, W. W. Lipids. (1996). Food chemistry, 1996, 225–320.

O'brien, R. D. (2008). Fats and oils: formulating and processing for applications. CRC press. 2008.

Okparanta, S. Daminabo, V. & Solomon, L. (2018). Assessment of rancidity and other physicochemical properties of edible oils (Mustard and corn oils) stored at room temperature. *J Food Nutr Sci.*, 6 (3), 70-5.

Osada, K.; Kodama, T.; Yamada, K.; Nakamura, S. & Sugano. K. Dietary oxidized cholesterol modulates cholesterol metabolism and linoleic acid desaturation in rats, fed high-cholesterol diets. Lipids, 1998, **33**, 757–764.

Özdemir, M. & Devres, O. (1999b). Turkish hazelnuts: Properties and effect of microbiological and chemical changes on quality. *Food Rev. Int.*, 15, 309 – 333.

Ozdemir, M. & Floros, J. *Crit. Rev. Food Sci. Nutr.*, 2004, 44, 185-193.

Pokorný, J.; Parkányiová, L.; Réblová, Z.; Trojáková, L.; Sakurai, H. (2003). Changes on storage of peanut oils containing high levels of tocopherols and β –carotene. *Czech J. Food Sci.*, 21, 19–27.

Proschogo, N. W.; Albertson, P.L.; Bursle, J.; Mcconchie, C. A.; Turner, A. G. & Willett, G. D. (2012). Aging effects on macadamia nut oil studied by electrospray ionization fourier transform ion cyclotron resonance mass spectrometry. *J. Agric. Food Chem.*, 60, 1973 – 1980.

Pryor, W. A.; Stanley, J. P. & Blair, E. (1976). 'Autoxidation of polyunsaturated fatty acids II: A suggested mechanism for the formation of TBA-reactive materials from prostaglandin-like endoperoxides', Lipids, 11, 370–379.

Ribeiro, M. A. A.; Regitano-D'arce, M. A. B., Lima, U. A. & Nogueira, M. C. S. (1993). Storage of canned shelled Brazil nuts (Bertholletia Excelsa): Effects on the quality. *Acta Aliment.*, 22, 295–303.

Robards, K., Kerr, A. F. & Patsalides, E. (1988). Rancidity and its measurement in edible oils and snack foods. A review. Analyst, 113(2), 213-224.

Rossell, J. B. 'Measurement of rancidity' in Allen J. C. and Hamilton R. J., Rancidity in Foods, London, Blackie Academic & Professional, 1994, 22–53.

Sacharow, S. (2006). *ActaHorta.*, 709, 125-126.

Sahasrabudhe, M. R & Farneig. (1964). 'Effect of heat on triglycerides of corn oil', *J Am Oil Chem Soc.*, 41, 264–267.

Sanders, T. A. B. (2010). Nutritional aspects of rancidity. 1989, *in*: Rancidity in Foods. Elsevier, 125–139.

Sattar, A.; Jan, M.; Ahmad, A., Hussain, A. & Khan, I. (1989). Light induced oxidation of nut oils. *Nahrung.*, 33, 213 - 215.

Shipe, W. F. & Senyk, G. F. (1981). Effects of processing conditions on lipolysis in milk. *Journal of Dairy Science.*, 64(11), 2146-2149.

Sonntag, N. O. V. (1976b). Reactions of fats and fatty acids. Bailey's industrial oil and fat products, 1(4), 499.

St. Angelo, A. J.; Vercellotti, J.; Jacks, T. & Legendre, M. (1996). Lipid oxidation in foods. *Critical Reviews in Food Science & Nutrition.*, 36(3), 175-224.

Stevenson, S. G.; Vaisey-Genser M & Eskin, N. A. M. (1984). 'Quality-control in the use of deep frying oils'. *J Am Oil ChemSo.*, 61, 1102–1108.

Swern, D. (1961). 'Primary products of olefinicautoxidations'. Autoxidation and Antioxidants. Interscience Publishers, 1961, I, 1–54.

Velasco, J.; Dobarganes, C. & Màrquez-Ruiz, G. (2010). Oxidative rancidity in foods and food quality. Chemical deterioration and physical instability of food and beverages. Woodhead Publishing, 3–32.

Virgili F.; Scaccini, C.; Packer, L. & Rimbach, G. (2001). Cardiovascular disease and nutritional phenolics. *in*: Antioxidants in Food. Woodhead Publ, Cambridge, 87–99.

W Chaiyasit, R. J, Elias, D. J, McClements & Decker, E. A. (2007). *Crit. Rev. Food Sci. Nutr.*, 47, 299-317.

Warner, K. & Frankel, E. N. (1985). Flavour stability of soybean oil based on induction periods for the formation of volatile compounds by gas chromatography. *Journal of the American Oil Chemists' Society.*, 62(1), 100-103.

White, P. J. (1995). Conjugated diene, anisidine value, and carbonyl value analyses. Methods to assess quality and stability of oils and fat-containing foods, 159-178.

Wsowicz, E.; Gramza, A.; Hêœ, M.; Jeleñ, H. H.; Korczak, J.; Maecka, M. & Zawirska-Wojtasiak, R. (2004). Oxidation of lipids in food. *Pol J Food Nutr Sci.*, 13(54), 87-100.

Printed in the United States
by Baker & Taylor Publisher Services